THE KING OF INFINITE SPACE

THE
KING *OF* INFINITE
SPACE

———◦◦◦———

EUCLID
AND HIS
ELEMENTS

———◦◦◦———

DAVID BERLINSKI

BASIC BOOKS

A Member of the Perseus Books Group
New York

Published by Basic Books,
A Member of the Perseus Books Group

Books published by Basic Books are available at special discounts for bulk purchases in the United States by corporations, institutions, and other organizations. For more information, please contact the Special Markets Department at the Perseus Books Group, 2300 Chestnut Street, Suite 200, Philadelphia, PA 19103, or call (800) 810-4145, ext. 5000, or e-mail special.markets@perseusbooks.com.

Designed by Trish Wilkinson
Set in 11.5 point Minion Pro

Library of Congress Cataloging-in-Publication Data
Berlinski, David, 1942–
The king of infinite space : Euclid and his Elements / David Berlinski.
   pages cm
   Includes index.
   ISBN 978-0-465-01481-1 (hardcover : alk. paper)—
   ISBN 978-0-465-03346-1 (e-book) 1. Mathematics, Greek.
   2. Geometry—History. 3. Euclid. 4. Euclid. Elements. I. Title.
   II. Title: Euclid and his Elements.
   QA31.B47 2013
   516.2—dc23
                                                        2012042492

10 9 8 7 6 5 4 3 2 1

For Morris Salkoff

On peut avoir trois principaux objets dans l'étude de la vérité: l'un, de la découvrir quand on la cherche; l'autre, de la démontrer quand on la possède; le dernier, de la discerner d'avec le faux quand on l'examine.

—BLAISE PASCAL, *DE L'ESPRIT GÉOMÉTRIQUE*

Contents

Preface   xi

I   Signs of Men   1
II   An Abstraction from the Gabble   11
III   Common Beliefs   19
IV   Darker by Definition   33
V   The Axioms   45
VI   The Greater Euclid   57
VII   Visible and Invisible Proof   77
VIII   The Devil's Offer   91
IX   The Euclidean Joint Stock Company   117
X   Euclid the Great   147

Teacher's Note   155
A Note on Sources   157
Appendix: Euclid's Definitions   159
Index   163

Preface

EUCLID IS UNIVERSALLY acclaimed great. His name is in no danger of being lost. He belongs in the company of men whose reputation defies revision. This is to establish Euclid's place, but hardly to say why so many years after his death, he continues to enjoy it.

Euclid is, of course, the author of the *Elements*, and the *Elements* is by far and away the most successful of mathematical textbooks. A textbook that has survived for more than two thousand years represents an uncommon achievement. Most textbooks have a short and ignominious life. They serve a purpose, but they do not inspire reverence.

Euclid's *Elements* is different.

No one has ever found a better way of presenting the elements of plane geometry; no sensitive teacher would think to use a substitute. There is none.

The *Elements* is not simply a great book in mathematics: it is a great book. The contemporary reader, eager for Euclid's personal revelations, will quit the *Elements* unsatisfied.

There is not a word about them. But in writing the *Elements*, Euclid found a way to impose his own powerful personality on the scattered propositions of geometry, and by imposing on them, created an immense structure, a logical space, a world in which there is growth, and form, and intimate dependencies among parts, something very large but not sprawling, the *Elements* itself the overflow in print, paper, or papyrus of a mind singular and determined.

Having revealed nothing of interest, Euclid, of course, has revealed everything of importance.

If this is not an artistic achievement, then nothing is.

Paris, 2012

THE KING OF INFINITE SPACE

Chapter I

SIGNS OF MEN

L'homme c'est rien; l'oeuvre c'est tout
(The man is nothing; the work, everything).

—Gustave Flaubert

THE ROMAN ARCHITECT Marcus Vitruvius Pollio lived and worked in the first century BC. His treatise *Libri Decem*, or *The Ten Books*, was dedicated to Caesar Augustus some twenty years before the birth of Christ. Vitruvius was both an architect and a military engineer, and *The Ten Books* contain a remarkable account of classical architectural ideas and building methods. It is sophisticated. A building, he insists, must be durable, useful, and beautiful (*firmitas, utilitas, venustas*). These are simple but stringent standards. Very few buildings constructed in the past sixty years could meet them. Vitruvius writes as a critic as well as a commentator, a man prepared to judge men as well as buildings, and when he does, he takes pride in seeing things as they are.

In his sixth book, *De Architectura*, Vitruvius recounts a story, one told as well by Cicero, about Aristippus, a

1

fourth-century philosopher. Finding himself "shipwrecked and cast on the Rhodian shore," he despaired.

Aristippus then happened to notice some geometrical figures scratched into the sand—triangles, perhaps, or circles, or straight lines suspended between points, the careless detritus of someone squatting by the seashore and thinking about shapes in space.

He said to his companions, "We can hope for the best, for I see signs of men."

Aristippus was well known for his devotion to pleasure; he was notorious. When rebuked for sleeping with whores, he responded equably that a mansion does not become useless because it has already been used. We expect such men to be tested; we are disappointed if they are not. It is right that Aristippus found redemption in human solidarity—*the signs of men.*

MATHEMATICS IS WHAT mathematicians make of it. What other standard would apply? Still, mathematicians exhibit a very nice sense of what they should make of what they have made. They are, after all, border guards at their own frontiers. Is mathematical logic a part of mathematics? Or mathematical physics? Most mathematicians would say they are not. They never doubt the importance of these subjects. They are not blind. But mathematicians are as fussy as cats. And almost as conservative. Their deepest commitment is to shapes and numbers, the seeing eye, the beating heart.

Counting would seem to come first, no? Every living creature makes the distinction between the thing that it *is* and the thing that it is *not*. Two numbers are required to express all of the imperatives of biology.

This is me, *that* is not.

Long live the numbers then.

But then there is seeing. Shapes are metaphysically as compelling as numbers. A single point, after all, divides the universe into what is *at* the point and what is *not*.

Long live the shapes too.

Shapes and numbers are in some sense coordinated. Points often have a numerical address. The latitude and longitude of Adelaide is 34 by 55S and 138 by 36E. The letters S and E may themselves be replaced by the numbers 19 and 5, their position in the alphabet. The result is a number marking Adelaide 345519138365. In just the same way, numbers often have a location. The number 345519138365 is notable in indicating the point where one finds *Adelaide*.

Long live the shapes *and* the numbers.

SOME CULTURES ARE geometric in their sensibility, and others are not.

Prizing order, the Romans of the empire appreciated severity. They did not fool around.

A powerful visual orthodoxy dominated their landscape: amphitheaters, public monuments and squares, cities divided into blocks, senatorial villas arranged in a rectangle around an interior space, a great urban civilization spreading throughout southern Europe and the Mediterranean basin.

Strange in a people whose numerals (I, II, XXXII) left them unable elegantly to conduct their practical affairs.

Our own culture is very different. The historian Tony Judt has argued that in the nineteenth century, the railroad, by shrinking space, brought about a reorganization of time.[1] A new standard of precision was first conceived and then enforced. Approximations that had long served the human race—sunrise, sunset, noon, midnight—were replaced by a complicated numerical apparatus, time divided into parts and parts of parts.

The result has been a culture that in comparison to the ancient world is numerically sophisticated but visually disgusting.

We count, *they* saw.

It makes a difference—obviously so.

---

1. Tony Judt, "The Glory of the Rails," *New York Review of Books*, December 23, 2010.

EUCLID OF ALEXANDRIA was born sometime in the fourth century BC, and he died sometime in the third. The year 300 BC is very often designated as a time in which he flourished—Euclid of Alexandria, 300 *fl.*, as historians sometimes write. Whatever the uncertainties of his birth and death, he was then at the height of his powers—alert, vibrant, and commanding. As a young man, Euclid is said to have been influenced by Plato's students, and he may well have attended the academy that Plato founded, mingling with the philosophers and inserting himself gregariously in their gathering gossip. Plato was devoted to geometry, even going so far as to assign to various deities a fondness for its study.

The circumstances under which Euclid composed his masterpiece, the *Elements*, remain, like the details of his life, largely unknown. There is some evidence that Euclid taught at the great library in Alexandria founded by Ptolemy I. The Euclid of the *Elements* is stern, logical, unrelenting, a man able to concentrate the powers of his mind on what is abstract and remote. It would be fascinating to know the details of his life in Alexandria, to *see* Euclid toddling off to the baths or with a sense that he has let things get out of hand, submitting to having his eyebrows trimmed. There are suggestions here and there that as a teacher, Euclid was urbane, helpful, and kind. Among its other virtues, the *Elements* is a great textbook; perhaps Euclid read aloud from his masterpiece as the warm sunlit air slitted through the dancing motes of dust, his students unaware that they were the first to hear a lesson that others would hear so many times and from so many other voices.

As a mathematician, Euclid took from his predecessors, men such as Eudoxus and Theaetetus, and gave to his successors, Apollonius and Archimedes. He summarized; he adjusted and refined; he was a living synthetic force and in very short order a monument—all this we know from what we can guess and from later commentaries, but the man himself remains invisible, his influence conveyed by his industry, a magnificent mole in the history of thought, a great digger of tunnels.

He must have been a man of heavy architecture, and at some point in his concourse with those endlessly gabbling philosophers, he gathered up his robes and, with a dawning sense of his powers, determined that he had something to offer that they had not seen and could not express.

For more than two thousand years, geometry has meant Euclidean geometry, and Euclidean geometry, Euclid's *Elements*. It is the oldest complete text in the Western mathematical tradition and the most influential of its textbooks.

The first book of Euclid's *Elements* contains 48 propositions, the second, 14. There are in all thirteen books comprising 467 propositions, and two more books of uncertain authorship, which are donkey-tailed onto older editions of the *Elements*.

The propositions in Books I through IV are concerned with points, straight lines, circles, squares, right angles, triangles, and rectangles, the stable shapes of art and architecture. Books V through IX develop a theory of magnitudes, proportions, and numbers. The remaining books are devoted to solid geometry. Every book of the *Elements* is compelling, but the Euclid of myth and memory is the Euclid of the first four books of his treatise.

In every generation, a few students have found themselves ravished by the *Elements*. "At the age of eleven," Bertrand Russell recalls in his autobiography, "I began Euclid, with my brother as my tutor. This was one of the great events of my life, as dazzling as first love. I had not imagined that there was anything so delicious in the world."

A course in Euclidean geometry has long been a part of the universal curriculum of mankind. Those not ravished by its study have very often remarked that the Euclidean discipline did them good nevertheless. It improved their mental hygiene. Students study algebra at roughly the same time that they study geometry, and curiously enough, they rarely remark on the improvement that it confers.

Algebra, students complain, is just nasty.

NOTHING FROM EUCLID'S hand survives into the twenty-first century. We know Euclid only from copies of copies, these passing through the mangler of translation from Greek to Latin and then to Arabic and back to Greek and finally to medieval Latin. Modern versions of Euclid are based on a tenth-century Greek manuscript identified in the eighteenth century by the French scholar François Peyrard. There is a poignant distinction between the solidity of Euclid's thoughts and the perishable papyrus he used to express them. Long before Euclid, the Babylonians wrote laboriously on tablets. *Plop* went the wet clay on a long table. Inscriptions by means of a curved stylus—*chiff, chiff, chiff*. The oven of the sun. And thereafter, immortality. We can see their words as well as their works. Euclid himself we cannot see at all.

If Euclid imposed order on his subject by making it a system, it was an order so severe as to force geometry into a fixed shape until at least the Italian Renaissance in the sixteenth century. Thereafter, a long and confusing process followed in which the Euclidean monument was variously chipped away, until in the nineteenth century, mathematicians discovered *non*-Euclidean geometries, Euclidean geometry becoming one among many, mathematicians half-mad with possibilities absorbing themselves with spaces that bulged like basketballs, or curved like saddle-backs, or that went on forever without getting anywhere.

Euclid's *Elements* represents the great achievement of the Greek mathematical tradition. Archimedes was a more brilliant mathematician than Euclid. He gave to the world what great mathematicians always give, and that is a record of his genius, but in the idea of an axiomatic system, Euclid gave to mathematics something even more enduring, and that was a way of life. It was a way of life invisible to the people who preceded the Greeks, and it was invisible as well to the Chinese, the masters of a subtle technological culture.

And as one might expect, it is invisible to everyone else as well—now, today, even so—and must as a result be taught like any other artifact of civilization.

Chapter II

AN ABSTRACTION
FROM THE GABBLE

*As all suns smolder in a single sun
The word is many but the word is one.*

—G. K. CHESTERTON

AN AXIOMATIC SYSTEM is a stylized organization of intellectual life, an abstraction from the gabble. Euclid conceived of an axiomatic system in order to fulfill an ambition that had before Euclid gone unconceived and so unexpressed: to derive all of the propositions about geometry from a handful of assumptions. The Egyptians who built the pyramids surely knew something about pyramids. They were not unsophisticated. They had a feel for measurements and mensuration. But what they knew, they knew incompletely. They took what they needed; they had no grasp of the whole. Euclid believed that there is a form of unity beneath the diversity of experience, and it is this that marks the difference between Euclid and the Egyptian mathematicians, men of the lash.

Euclid required a double insight before he could strike for immortality. The first: that the various propositions of geometry *could* be organized into a single structure; and the second: that the principle of organization binding geometric propositions *must* be logical, and so alien to geometry itself.

These are radically counterintuitive ideas, Pharaonic in their audacity.

Euclid's assumptions are commonly called axioms, and sometimes postulates; his conclusions, theorems. A proof is a chain connecting the axioms to the theorems in logically unassailable links. Euclid assumed five axioms, and from these he derived 467 theorems.

A sense of this intellectual power, its grandeur—this is the Euclidean gift. The Pythagoreans before Euclid were men consumed by the rapture of mathematics. They communed with the numbers, and they were often tempted by gross intellectual follies. They took pleasure in mumbo jumbo. Euclid is by comparison imperturbable. There is no rapture in the *Elements*, but neither is there anything insane. The structure that Euclid created is intellectually accessible to anyone capable of following an argument.

Like the pyramids, an axiomatic system is a public work.

EUCLIDEAN GEOMETRY IS the study of shapes in space. Shapes are not bound to the wheel of time. There is no place where the Euclidean triangle resides, and no time at which it arrived there. Plato argued that the shapes are a part of the Kingdom of Forms, caves and cavemen, shadows and the ecstatic sun. No philosopher since Plato has been entirely satisfied with the kingdom. Existing in the great beyond, the Platonic forms have no obvious causal powers. Yet if they have no obvious causal powers, they have obvious causal effects. Euclid reached his conclusions about triangles by reasoning about the form of the triangle, the essential thing.

If the Platonic forms are difficult to accept, they are impossible to avoid. There is no escaping them. Mathematicians often draw a distinction between concrete and abstract models of Euclidean geometry. In the abstract models of Euclidean geometry, shapes enjoy a pure Platonic existence. The concrete models are in the physical world. Freeways masquerade as straight lines, ink drops as points, amphitheaters as circles, and planetary orbits as ellipses. Mathematicians have often supposed that the concrete models of Euclidean geometry have a degree of vitality denied the Platonic models. "One must be able to say at all times," the German mathematician David Hilbert remarked, "instead of points, straight lines, and planes—tables, beer mugs, and chairs." These words convey a reassuring impression of ordinary life. Tables, beer mugs, and chairs! What could be more down-to-earth? But the phrase *instead of* prompts a reservation. The shortest distance between two beers is a straight line in time *or* in space. Yes, that is certainly true. But the shortest distance between two beers is a straight line *because* the shortest distance between two points is a straight line. Nothing is instead of anything.

Without the Platonic models, the concrete models would have no interest. Euclid does not, after all, invite his readers to consider *more or less* straight lines. How much more, how much less? And if there are no purely straight lines, what would the comparison come to?

The concrete models of Euclidean geometry include the tables, chairs, and beer mugs. They are where they have always been, and that is in the barroom.

The Platonic models of Euclidean geometry include the points, lines, and planes. They are where they have always been as well, and that is God knows where.

If the theorems of an axiomatic system follow from its axioms, it is reasonable to ask what *following from* might mean. What *does* it mean? The image is physical, as when a bruise follows a blow, but the connection is metaphorical. The relationship between the axioms and the theorems of an axiomatic system is, when metaphors are discarded, remarkably recondite, invisible for this reason to all of the ancient civilizations but the Greek.

The men of the ancient Near East no doubt knew what arguments were. They had so many of them. What they knew, they knew imperfectly. They lacked words to make clear the distinctions that they sensed. Why assess an argument when it was so much easier to end it by either violence or indifference? This point of view has never completely lost favor. It was the Greeks who did the assessment and forced the very idea of an inference into consciousness, asking patiently for an account of its nature, the way it controlled the movement of the mind, and where in the catalog of human powers it belonged.

At roughly the same time that Euclid composed the *Elements*, Aristotle provided a subtle and refined analysis of syllogistic inference, the pattern in argument that takes Socrates—and the rest of us, alas—to his death by virtue of the fact that he is a man and we are mortal. Born in 384 BC and dying in 322 BC (another victim of his own syllogism), Aristotle might conceivably have known Euclid when Euclid was still a young man, perhaps even palpating his togaed shoulder. Far too little is known about the circumstances of Euclid's life to say just whose hand he might have shaken. The two men worked hand in hand all the same.

ARGUMENTS, ARISTOTLE ARGUED, may be divided into those that are good and those that are not. In the syllogism, two premises resolve themselves in one conclusion:

All dogs are mammals.
All mammals are animals.
All dogs are animals.

Good.

No fish are dogs.
No dogs can fly.
All fish can fly.

Bad.

Any dog who has not lost something still has it.
No dog has lost a fifth foot.
All dogs have five feet.

Shame.

As these examples might indicate, good arguments are good by virtue of their form and not their content. The logician is indifferent to the distinction between *all dogs are mammals* and *all men are mortal*; both cases are swallowed whole by *all A's are B*. This is the Aristotelian insight, and logicians have accepted it ever since. The conclusion of a valid argument is entrained by its premises. Truth plays an ancillary role. If the premises of a valid argument are true, then their conclusion must be true, but *whether* they are true is a matter on which the logician has little to say; an argument may be good even though its premises are false, and bad even though its premises are true.

It is tempting to imagine a fraternal give-and-take between Euclid and Aristotle, Euclid taking, Aristotle giving, with Euclid advancing proofs and arguments that Aristotle had antecedently assessed and classified. This is not quite so. The *Elements* is a work of great logical sophistication, but it is not a work of logical self-consciousness. Euclid's subject is geometry, his business is proof, and Euclid was not a mathematician disposed to step back to catch himself in the act of stepping back. That his arguments were valid, he had no doubt, but in the question of what made them valid, he had no interest.

Euclid often made use of arguments that Aristotle had not analyzed properly or analyzed at all. *If* the natural numbers progress by 1, *then* there is no natural number between 3 and 4. The natural numbers *do* progress by 1. So there is no natural number between 3 and 4. The inference proceeds by the stately music of *modus ponens*. No syllogism is involved, just the straightforward play between propositions and their particles—*if, then, and*. Euclid is especially fond of reaching his conclusions by demonstrating that a given proposition leads to a contradiction, and so must be rejected. In Euclid's hands, this style of reasoning often becomes a torpedo.

There remains the matter of the distinction between an axiomatic system and an argument.

There is none.

An argument is a small axiomatic system, and an axiomatic system is a large argument.

Chapter III

COMMON BELIEFS

> La dernière démarche de la raison, c'est de connaître qu'il y a une infinité de choses qui la surpassent. *(The last step of reason is to grasp that there are infinitely many things beyond reason).*
>
> —PASCAL

"EUCLID ALONE," EDNA St. Vincent Millay once wrote, "has looked on beauty bare." It is the first line in a sonnet of the same name. Literary critics are often embarrassed by the sonnet, and mathematicians by Edna St. Vincent Millay. Euclid *alone*? Still, the idea that "Euclid alone looked on beauty bare" elegantly draws attention to the nakedness of inference exhibited by every Euclidean proof. It is something rarely seen beyond mathematics—this hidden, if somewhat lurid, power of a Euclidean proof to compel fascination. Up go the axioms on the blackboard; down come the theorems. Students and readers alike are encouraged to think of the display as something stirring.

And it is. So much of ordinary argument and inference is fully clothed.

But this way of presenting Euclid and the *Elements* imposes a gross distortion on Euclid's thoughts: it allows the staged drama of his proofs to stand for the grandeur of his system as a whole. Euclid meant his proofs to be grasped against the background of his common notions and definitions. In almost every proof, he appeals to his own common notions and, in many proofs, either to his definitions or to ideas that follow naturally from his definitions. Beyond any of this, there are Euclid's ideas about space and human agency and the exaltation of geometry that is so conspicuous a feature of his thoughts. Focus, control, and tension—they are there in Euclid's proofs, but these moments, as any athlete knows, do not appear as isolated, brief, bursting miracles. They are not isolated at all, and they are not miracles either. They are grounded in Euclid's meditations about what may be supposed and what not, and how difficult ideas may be defined or, at least, exposed. In all this, the master, unbending to explain himself, remains entirely in character, his severity undiminished, no word wasted, as prudent, compact, and tight as the stretched skins on which he wrote.

Euclid's common notions represent the "beliefs on which all men base their proofs." The words are Aristotle's, but the idea that there *are* beliefs on which all men base their proofs must itself have been one of them, for Euclid appropriated the idea without hesitation and without argument.

There are five common notions in all:

1. Things that are equal to the same thing are also equal to one another.
2. If equals be added to equals, the wholes are equal.
3. If equals be subtracted from equals, the remainders are equal.
4. Things that coincide with one another are equal to one another.
5. The whole is greater than the part.

These principles convey an air of what is obvious. They have authority. No one either in Euclid's time or our own is proposing that if equals are added to equals, the result might be *un*equal. A surprising delicacy is nonetheless required to say just what these principles mean. It is a delicacy that Euclid did not possess. This might suggest that Euclid's conviction that these beliefs are *common* represented on his part a willingness to repose his confidence in things he could neither explain nor justify. To say as much involves no rebuke. If Euclid could neither explain nor justify the common beliefs that he invoked, we can do as little, or as much, with respect to our own. It was Euclid's genius to grasp that whatever the powers of his geometrical system, it *did* rest on certain common beliefs.

It was Euclid's business to say what those beliefs were. And our business to say what they mean.

EQUALITY IS AN indispensable idea. It is like water to the fish—everywhere at once, but easy to ignore and difficult to define. To say of *two* things that they are equal is always false, and to say of *one* thing that it is equal to itself is always trivial. This is an uncommonly stern conceptual rebuke.

Euclid's first common notion is often illustrated by three straight lines labeled A, B, and C and an insouciant appeal to intuition. If A is equal to B, and B is equal to C, then A is equal to C.

The appeal is not misplaced, but it is misleading. For one thing, neither illustration nor intuition says much about the concept of equality. For another thing, what Euclid says of equality is also true of size: if A is greater than B, and B greater than C, then A is greater than C.

Euclid's statement of his first common notion covers up a *chamboulement*, a disorder. The illustration, those lines—this is starting well. But two equal lines? With the long history of Euclidean geometry at our back, we can say easily enough that two lines are equal if they are equal in *length*. A one-foot line in Moscow is the same length as a one-foot line in Seattle. But equality in length is a far narrower concept than equality itself, and it is not a concept that Euclid made accessible to himself. Euclidean geometry contains no scheme under which numbers are directly associated with distances.

Euclid's fourth common notion expresses the Euclidean concept of geometrical equality. Having been in the grapple, Euclid has, we may suppose, gotten the better of things. Two things are equal if they coincide. This principle of superposition Euclid puts to work throughout the *Elements*. In the case of those straight lines, it admits of immediate application. Two lines are equal if they coincide. A question having been posed about equality, a very similar question now arises about coincidence: just when do things coincide? To say that two things coincide when they coincide equally is not obviously an improvement. Having fastened on coincidence as crucial, Euclid may well have remembered that in his definitions, he affirms that a line, although it has length, has no width. What investigation might justify the conclusion that two lines without width coincide? If no investigation, how could we say that two lines coincide even in length if we cannot say whether they coincide at all?

The wheel of time required twenty-three centuries before George Boole and C. S. Peirce assessed equality in its proper, its logical, context. Mathematicians today take it all in stride. Aristotle and Euclid were more strode upon than striding.

THE PROPOSITION THAT Euclid is wise says of Euclid that he is wise. His wisdom is something that he has, an aspect of the man. *Euclid is wiser than Aristotle* says of Euclid and Aristotle that one man is wiser than the other. It puts them both in their places—two men, but one relationship.

Equality is a relationship and, as such, a member of a great, worldwide fraternity: things bigger, taller, slighter, smaller, greater, grander, fathers and sons, daughters and mothers, before and after. To them, the logic of relationships, a general account of just how an A might be related to a B, the rules of the road.

Equality is in the first place *reflexive*. A = A. No relationship could be closer. Or more universally enjoyed.

And *symmetric*. If A = B, then B = A.

And *transitive*. If A = B, and B = C, then A = C.

Euclid saw the transitivity of equality. It is the first of his common notions. But symmetry and reflexivity he missed or did not mention.

In his second and third common notions, Euclid juxtaposes the relationship of equality and the operations of addition and subtraction. Things are added to one another or subtracted from one another. Inasmuch as subtraction is a way of undoing addition, Euclid's second and third common notions might be collapsed into one encompassing declaration: If A = B and C = D, then A ± C = B ± D.

There is no reason, one might think, to restrict this principle to arithmetical operations; there is no reason to restrict it at all. A = B if and only if whatever is true of A is true of B. This is sometimes thought a definition of equality, and so a way of eliminating a troublesome concept altogether. It is not clear that this maneuver confers any great benefits. Among the things true of A is surely that A is equal to itself. The concept destined to be disappeared has just been reappeared. This might suggest that equality cannot easily be eliminated in favor of the truth because it cannot be eliminated at all.

Just so.

EUCLID'S FOURTH COMMON notion expresses a criterion of identity, a principle by which triangles, circles, or straight lines may be judged the same. The idea is implicit in every theorem that Euclid demonstrates. It is of the essence. If the geometer cannot tell when two shapes are the same, he cannot tell when they are different, and if he cannot tell whether shapes are the same or different, of what use is he?

Suppose now that two triangles are separated in space. They become coincident when one of them is moved so that it covers the other in such a way that the two figures are perfectly aligned. Nothing is left over, extrudes, or sticks out.

Coincidence or superposition offers the geometer a rough-and-ready measure of sameness in shape. What is not entirely obvious in all this rough-and-readiness is just how figures separated in space—a triangle here, another one there—can be moved through space so that their coincidence may be tested.

The point emerges early in the *Elements*; it emerges in Euclid's fourth proposition:

> If two triangles have the two sides equal to the two sides respectively, and have the angles contained by the equal straight lines equal, they will also have the base equal to the base, the triangle will be equal to the triangle, and the remaining angles will be equal to the remaining angles respectively, namely those which the equal sides subtend.

Two triangles are equal, Euclid has affirmed, if they are congruent, and they are congruent if two of their sides are equal, along with the angles the sides subtend.

The proof is simple in its notoriety, for Euclid deviates at once into the swamp of concepts that he has not analyzed and cannot justify: "If," he says, "the triangle ABC be applied to the triangle DEF, and if the point A be placed on the point D and the straight line AB on DE, then the point B will coincide with the point E, because AB is equal to DE."

Euclid is at the podium. He has just pointed to his dust board with the tip of an outstretched finger. Beaming with satisfaction, he is about to say . . .

When he is interrupted.

—Applied by whom, Sir?

One question.

—Placed how, Professor?

Another.

—Coincide when, *Maître*?

A third.

BOTH BERTRAND RUSSELL and David Hilbert thought that Euclid would have been better served had he accepted proposition four as an axiom instead of claiming it as a theorem. It is a policy, as Russell remarked in another context, that has all the advantages of theft over honest toil. Designating Euclid's fourth proposition an axiom does not do much to diminish the sense that in moving things around on the blackboard, the geometer has undertaken something at odds with the rigor of Euclidean geometry. In a little book titled *Leçons de géométrie élémentaire* (Lessons of elementary geometry), the French mathematician Jacques Hadamard proposed that coincidence be subordinated to some catalog of the ways in which shapes in Euclidean space might move. If the Euclidean idea of coincidence is a theorem, it depends on assumptions that Euclid did not make; if an axiom, it makes those assumptions without defending them; and if based on some antecedent assessment of motions allowed Euclidean figures, then it is both.

The distinction between the concrete and the abstract models of Euclidean geometry offers a nice place in which to watch this uneasiness emerge and then separate itself into a destructive dilemma.

Does the idea of coincidence apply to the concrete or the abstract models of Euclidean geometry? Or neither, or both? Not the concrete models, surely, for physical triangles are never completely coincident, no matter how they are moved. Something is always left out, or something always left over. How on earth can two physical objects coincide perfectly?

Not on earth is the correct answer; it is the only answer. If it is true that concrete triangles are never coincident, it is equally true that abstract triangles cannot be moved. They are beyond space and time. Moving about is not among the things that they do, because they do not *do* anything.

Sensitive to just this point, Russell dismissed the idea that in Euclidean geometry, anything is moving or being moved. Writing in the supplement to the 1902 edition of the *Encyclopedia Britannica*, Russell remarked that "what in geometry is called a motion is merely the transference of our attention from one figure to another."

But the geometer's attention is like the wind: it goeth where it listeth. Where it goeth is of little interest unless it goeth from one figure to another equal figure.

Coincidence is a condition that the concrete models of Euclidean geometry cannot satisfy: they are never the same.

And it is a condition that the abstract models of Euclidean do not meet: they cannot be moved.

THERE IS FINALLY the last of Euclid's common notions, the principle that the whole is greater than the part. Far from expressing a belief on which "all men base their proofs," the proposition is either trivially true or false.

If the whole of something is by definition greater than its parts, Euclid has not advanced his cause or his case; but if the very idea of a part standing to a whole is left undefined, it is easy enough to construct examples in which the whole is less than its parts or equal to them.

The number 6, to take an example, has its own internal structure. It may make sense to say that 0 and 1 are simple numbers, quite without parts, but the number 6 is the sum and product of various numbers and thus has a richness in its identity, an otherwise hidden complexity. Is the number 6 greater than its parts? Is it greater than the *sum* of its parts? Not if the parts of the number are composed of its divisors, 1, 2, and 3. Their sum is equal to 6.

The number 12, on the other hand, is *less* than the sum of its parts, 1, 2, 3, 4, and 6.

The relationship between wholes and parts is exquisitely sensitive, then, to the way in which the underlying ideas are specified. If this is so, then it is difficult to ascribe Euclid's fifth common notion to those beliefs "on which all men base their proofs." Too much by way of circumstantial dependency is involved for this to be a common notion at all.

Infinitely large objects present problems all their own. Is the assertion that the whole is greater than its parts true of the natural numbers? Skepticism arises because the natural numbers 1, 2, 3, . . . may be put into a tight correspondence with the even numbers 2, 4, 6. . . . The correspondence is tight enough so that for every natural number, there is an even number, and vice versa. The set of natural numbers and the set of even numbers, as logicians say, have the same cardinality. They are the same size.

But surely, the even numbers are a part of the natural numbers? If they are not, what residual meaning can be assigned to the now-vagrant terms *part* and *whole*?

The goal of listing once and for all those ideas on which "all men base their proofs" is profoundly compelling. A list is something explicit and thus open to inspection; once open to inspection, a list of common notions satisfies the desire to have all the cards on the table. Hidden assumptions, like hidden cards, suggest that what is hidden is somehow disreputable.

The explicitness with which Euclid affirms certain common notions is, of course, no reason by itself to think his common notions any good. Euclid never suggests otherwise. His common notions are what they seem. They express assumptions that are more general than his axioms but no less undefended.

If Euclid's common assumptions cannot be derived from anything further, they make their claim by means of their inescapability. Without them, Euclid believes, there could be no proof at all. Whatever their inescapability, Euclid's common notions suggest a question that neither he nor Aristotle ever considered. Can these common notions be faulted because they are incomplete? Whenever an explicit list of common assumptions is offered, after all, it is easy enough to step back and with some assurance point to the assumptions on which the assumptions themselves depend.

Like any other mathematician, Euclid took a good deal for granted that he never noticed. In order to say anything at all, we must suppose the world stable enough so that some things stay the same, even as other things change. This idea of general stability is self-referential. In order to express what it says, one must assume what it means.

Euclid expressed himself in Greek; I am writing in English. Neither Euclid's Greek nor my English says of itself that it *is* Greek or English. It is hardly helpful to be told that a book is written in English if one must also be told that *written in English* is written in English. Whatever the language, its identification is a part of the background. This particular background must necessarily remain in the back, any effort to move it forward leading to an infinite regress, assurances requiring assurances in turn.

These examples suggest what is at work in any attempt to describe once and for all the beliefs "on which all men base their proofs." It suggests something about the ever-receding landscape of demonstration and so ratifies the fact that even the most impeccable of proofs is an artifact.

Chapter IV

DARKER BY DEFINITION

Sometimes things may be made darker by definition. I see a cow. I define her, Animal quadrupes ruminans cornutum. *Cow is plainer.*

—Samuel Johnson

THE *ELEMENTS* CONTAINS twenty-three definitions. Of these, the first seven, and the twenty-third, are fundamental:

1. A point is that which has no part.
2. A line is length without breadth.
3. The extremities of a line are points.
4. A straight line is a line which lies evenly with the points on itself.
5. A surface is that which has length and breadth only.
6. The extremities of a surface are lines.
7. A plane surface is a surface which lies evenly with the straight lines on itself.

23. Parallel straight lines are straight lines which, being in the same plane and being produced indefinitely in both directions, do not meet one another in either direction.

Nineteenth- and twentieth-century mathematicians have almost to a man objected to these definitions. Both Moritz Pasch and David Hilbert criticized Euclid because in struggling to say what he meant, Euclid rejected what he knew: things come to an end. If axioms must be accepted without proof, then some terms must be accepted without definition. In his ninth through twenty-second definitions, Euclid is almost impeccable, defining terms that are new by an appeal to terms that are old. A triangle is a figure contained by three straight lines. This is Euclid's nineteenth definition. Not perfect. What is a figure? But not bad. There remain his initial definitions. A point, Euclid affirms, has no parts. It is the first thing that he says, circumstances suggesting that he meant to say it. And since it is the first thing Euclid says, it is the first definition to which critics object. "This [definition] means little," Morris Kline argues in *Mathematical Thought from Ancient to Modern Times*, "for what is the meaning of parts?"

Yet if Kline intended to rebuke Euclid for a logical mistake, he has done so by making a mistake all his own. The haunch of a cow is one of its parts, but only the word *haunch* carries meaning. The haunches have nothing to do with it.

They are busy supporting cows. When physicists say that an electron has no parts, they are talking about electrons and not the meaning of the words that might denote them. So, too, Euclid. His first definition seems much less a proper definition than a fact about points: that they have no parts.

In his twenty-three definitions, Euclid blurs the line between the claims that he makes and the terms that he defines. He is not fully the master of distinctions. His definitions are, for this reason, moving. They reveal a great mind entering uncertainly into a space that logicians would not fully command for two thousand years. The definitions are what they seem: an instruction, a guide to Euclid's thoughts, a way into the labyrinth.

GEOMETRY IS THE study of shapes in space. There is not one without the other. The enveloping plane, Euclid makes clear in his fifth definition, has in length and breadth two dimensions. What is a dimension, and why are there two of them? Length and breadth are terms of ordinary experience— hands outstretched or one above the other, as if measuring a flounder. The third dimension of space is often represented by extending a flat palm forward—*in*—and then retracting it—*out*. This reversion to experience might suggest that the Euclidean plane is simply whatever is left over when one dimension of space is subtracted from the original three. It is a position impossible to fault.

Indifferent to sturdy common sense, textbook authors often say that the plane has two dimensions because two numbers are sufficient to identify any point. It is by no means clear that this is the improvement commonly supposed. Two numbers are sufficient to identify any point in space if the space has *two* dimensions. Not otherwise. If we had some analytic understanding of just how points comprise a space of two dimensions, there would be no need to appeal to two numbers, and if not, of what use is the appeal?

Euclid's introduction of length and breadth is nevertheless not entirely misplaced. An appeal to the power of geometrical objects to move, or to be moved, is latent throughout the *Elements*. It is the power behind proposition four, and thus the point of coordination to Euclid's system of equality, the one in which he considers shapes the same or different. It hardly matters whether a geometrical object, having taken it in its head to vacate its premises, moves on its own or is moved by the geometer. With motion assigned geometrical figures, in how many ways could any one of them move? The behavior of a marble on a plate-glass table suggests that there is no end to the possibilities. And this is true enough. There is no end. What the question demands, however, is not an overall count but a kind of classification, a reduction to essentials by which what the moving marble does may be collapsed into a finite scheme.

There are three ways to move in the plane: by *translation*, moving straight ahead on any convenient straight line: by *rotation*, turning in an arc at a point; and by *reflection*, as when a flounder's horrible two eyes, having looked out from the plane, are persuaded out of common decency to look into the plane.

This is the finite scheme.

THERE ARE THREE degrees of freedom in the Euclidean plane, geometers say—a nice phrase and a reminder that even in mathematics, there are ties between ideas that are austere and abstract—degrees—and ideas that appeal to human agency—freedom. If there are three degrees of freedom, then two dimensions. A little formula coordinates the degrees of freedom and dimensionality: $n (n + 1) / 2$, where n is the dimensions of space, and $n (n + 1) / 2$, the degrees of freedom.

One thing, the dimension of space, has been defined in terms of another, its degree of freedom, but there is a long, glistening rodent trail leading backward from these ideas to the far more primitive idea of some mental movement by which the *geometer* shuttles between dimensions, undertaking observations and seeing things where in real life no observations could be made and nothing could be seen.

EUCLIDEAN SPACE HAS a dash of the distant in its veins; this much is clear from Euclid's twenty-third definition. The word *infinite* does not appear in the definition itself. What Euclid does say is that straight lines may be produced indefinitely in both directions. There is as much of space, it follows, as might be needed to encompass an ever-expanding straight line. Still, the definition is peremptory, failing to make distinctions that are dying to be made. There is a difference, after all, between a space that is unbounded and one that is infinite. The surface of a sphere is unbounded but not infinite, and a line of fractions narrowing to zero is infinite but not unbounded. Just a century before Euclid wrote, Zeno the Eleatic provided a discussion of infinity and its paradoxes that is to this day matchless in its subtlety. It is possible that Euclid said as little as he did because he understood that he stood to gain nothing by saying more.

Why make trouble?

ABOUT FLATNESS, EUCLID has two things to say. A straight line, he affirms in his fourth definition, lies evenly with the points on itself, and in his seventh definition, Euclid makes much the same claim for the plane itself. The plane lies evenly with its embedded straight lines. The contrast is between straight lines and curves, and between the flat and level plane and other surfaces such as the surface of a sphere.

The idea of flatness has a certain emotional valence well beyond geometry. Ideas, champagne, and chests may all be flat; this is rarely said to be a good thing. Both falling flat and flattery share a common Old Norse root in *flatr*, which designates a leveling down, a featurelessness. The Euclidean plane is everywhere the same. This invites the question: The same with respect to *what*? In the concept of superposition, or coincidence of shape, Euclidean geometry makes a concession to the idea that Euclidean figures may be moved. The triangles of proposition four are congruent to themselves no matter how they shuffle or are shuffled across the plane. Euclidean figures are indifferent to the fallacy that distance makes a difference. *Caelum non animum mutant qui trans mare currunt*, as Horace observed. They change the sky but not their souls who flee across the sea.

But if the Euclidean plane is homogeneous, it hardly follows that it is flat. The sphere is everywhere the same, and so is the geometry of the earth, as jaded travelers well know. It is not flat.

IN THE CALCULUS, the curvature of a line is defined by an appeal to the straightness of straight lines; they have no curvature at all. In his treatise *Relativity and Geometry*, the physicist Roberto Torretti writes that "the curvature of a plane curve at a point measures the rate at which the curve is changing direction." Curvature is a falling away. Torretti then adds something wonderfully vivid. What curvature really measures at a point is the extent to which a curve is "departing from straightness."

Surfaces as well as curves may depart from straightness. If the plane were balanced on the top of a sphere, like a book balanced on an apple, then one might say that the sphere is curved at its apex, by virtue of the increasing distances between the plane and the surface of the sphere. The apple has undertaken its own departure.

To see this requires of an observer a complicated maneuver in which apple and book, plane and sphere, are somehow embedded in a three-dimensional space, the extra dimension required to place both objects in juxtaposition. The result is a standard measure of curvature and so of flatness—*extrinsic* curvature, to use the suggestive name given it by mathematicians—with curvature now a relative property, one space curved when measured by the standards of another, almost as if what is crooked could be understood only against what is straight. It is a principle known to be useful in the criminal law as well as in mathematical physics.

Still, there is no ultimate decisiveness to extrinsic curvature. The sphere is curved when measured against the plane. The first has positive curvature—it swells—and the second, no curvature at all. It is flat. But wherein flatness itself?

Is THERE A measure of flatness accessible to an observer within a two-dimensional space, to an ant, say? Could that ant, bound forever to wander the blackboard, *discover* that the blackboard is flat? The answer was provided by Carl Friedrich Gauss in a remarkable theorem that he published under the title *Theorema Egregium*. The intrinsic curvature of a surface, Gauss demonstrated, may be deduced entirely by using local clues such as angles and distances and the way that they change. No appeal to spaces beyond a surface is necessary, and what is more, intrinsic and extrinsic curvature coincide and they coincide perfectly.

In reaching these conclusions, Gauss went considerably beyond anything in Euclidean geometry itself. His *Theorema Egregium* is an exquisite achievement, but it is an exquisite achievement in differential geometry, one of the innumerable mixed marriages in mathematics, this one between the analytic apparatus of the differential calculus and the classical concerns of Euclidean geometry. Euclid did not discuss differential geometry and could not have foreseen its development.

So FAR, SO pretty good.

What lies between two points in the Euclidean plane? One answer is nothing. This is the answer suggested by Democritus in the fifth century BC. There are in nature only atoms and the void, Democritus argued, the atomic theory of matter just budding at his fingertips. Ancient atoms were both indivisible and indestructible. In the twenty-first century, those atoms have given way to elementary particles, but the idea of a radical dissection of material objects into their parts remains as imperishable as the atoms it countenances.

There is a very considerable difference between a physical atom and a Euclidean point, if only because one is physical, the other not, but Euclid in his study may well have felt Democritus behind his back, a gray ghost hanging over his shoulder, as ghosts so often do, one man's point an idealization of the other man's atom. Nothing between atoms; nothing between points; and, so, nothing all around.

By whatever means he found himself in Euclid's study, Democritus was not alone. Parmenides, his predecessor, was there, too, muttering. At some time in the fifth century, Parmenides had composed a long poem titled *On Nature*. Surviving in fragments, his voice comes to us over an immense distance, sun-baked, half-mad, delirious. It is not at all modern.

"What is, is," Parmenides says, and as for what is not, "it is not."

It is difficult to imagine an objection being framed. Did anyone in the fifth century BC propose that what is, is not, or that what is not, is? Yet from the premise that nothing is, after all, nothing, Parmenides drew the conclusion that there is no void between atoms, because it makes no sense to say of a void that it *is*.

It then follows that space is just one thing, and not many things. What beyond spatial separation could mark the distinction between atoms, the more so if like Euclidean points, they have no parts? If space is filled, then motion and, indeed, change, are impossible. There is no place to go, and if no place to go, no place to have come from either.

These strange ideas belong to the pre-Socratic world, one that in the popular imagination contrasts unfavorably with our own. But Euclid lived and worked within historical memory of the pre-Socratic philosophers. Parmenides was as close to his consciousness as Abraham Lincoln is to ours. Those bony Parmenidean fingers were poking into Euclid's shoulder.

If there are points in the plane, then Euclidean space is replete with them, for between two Euclidean points along any straight line, there is always another Euclidean point. The inference is almost immediate. Euclid's third definition identifies the ends of a line with two points, and his twenty-third definition establishes that a straight line may be produced indefinitely. Suppose that there is *no* point between the points P and Q lying on the straight line L. Then starting at P, L could not fall short of Q. Lacking any other point by assumption, one of its ends would dangle uselessly. In that case, how could L be produced from P?

This downward descent by which points lead to points must, so one might imagine, end either with nothing between points or with something still further. It is an inference at odds with the geometer, eager to get from one point to another.

If nothing, how? If something, what?

IN THE COMPETITION between contending ghosts, Parmenides has made his influence felt. Democritus, too. Euclidean points may well be like atoms, but there is no void anywhere in the *Elements*, no suggestion that there is nothing between points. For Euclid, it is points all the way down.

The discussion is hardly at an end. In his little book *Das Kontinuum*, the twentieth-century mathematician Hermann Weyl found himself interrogating the pre-Socratics all over again. It is quite a crowd in Euclid's study. Between any two points, there is a third. Yet time flows, and things change, and there is a distinction between the flow of time and the points used to mark that flow. The points are like diamonds in a skein of silk: attend to them, and they catch. But as time flows, it does not catch. "The view of a flow," Weyl wrote, "consisting of points and, therefore, also dissolving into points turns out to be mistaken: precisely what eludes us is the nature of the continuity, the flowing from point to point; in other words, the secret of how the continually enduring present can continually slip away into the receding past."

About these issues, Euclid said nothing at all.

Chapter V

THE AXIOMS

> Nempe nullas vias hominibus patere ad
> cognitionem certam veritatis praeter
> evidentem intuitum, et necassariam
> deductionem *(There are only two routes
> open to human beings to arrive at sound
> knowledge of the truth, evident intuition
> and necessary deduction).*
>
> —René Descartes

> *The dull mind, once arriving at an
> inference that flatters the desire, is rarely
> able to retain the impression that the
> notion from which the inference started
> was purely problematic.*
>
> —George Eliot

EUCLID PROPOSED FIVE axioms for geometry. These axioms cannot, of course, be themselves derived from still further assumptions. Or from anything else. "No science," Aristotle dryly remarks, "proves its own principles."

It is possible, of course, that if some theorems were made axioms, then some axioms could be made theorems. The American logician Harvey Friedman has for this reason studied the extent to which something standing on its feet could be made to stand on its head. This does not mean that Euclid's axioms are unjustified or arbitrary. If that were so, what would be their interest? Euclid accepted self-evidence as the justification for his axioms, and he was troubled to discover that not all of his assumptions were evident, not even to himself.

The first three of Euclid's axioms are commonly grouped together. "Let the following be postulated," Euclid writes:

1. To draw a straight line from any point to any point.
2. To produce a finite straight line continuously in a straight line.
3. To describe a circle with any center and distance.

These assertions are hardly controversial. They seem to make perfect sense. Two points, one straight line. What could be simpler? But if intellectually disarming, these axioms are also disconcerting. They cede to the reader powers properly the mathematician's, or if not the mathematician's, then obviously not the reader's: to *draw*, to *produce*, and to *describe*.

What if that reader, unwilling to do anything, is unwilling to draw, produce, or describe? Or if he does not know how? What then? "Geometry does not teach us to draw these lines," Isaac Newton remarked in the *Principia*, "but requires them to be drawn."

Euclidean geometers have traditionally explained the first three of Euclid's axioms by reference to a straight-edge and compass. In his wonderful companion to Euclidian geometry, *Geometry, Euclid and Beyond*, the contemporary mathematician Robin Hartshorne remarks that Euclid's proofs are "carried out with specific tools, the ruler (or straightedge) and compass." Faithful to his policy of saying as little as possible, Euclid himself never once mentions either a straight-edge or a compass in the *Elements*. Nor does Hartshorne. When at last he defines a geometrical construction, Hartshorne abjures both ruler and compass and writes instead about "constructible numbers."

Having been introduced at some moment after Euclid put down his stylus, the straight-edge and compass proved a very considerable success. Students enjoyed stabbing paper with a compass point and drawing aimless circles. Some things could be done with just these two instruments, and some things not. This made for a nice series of discoveries. It is impossible to square a circle using only straight-edge and compass, and impossible again to trisect an arbitrary

THE KING OF INFINITE SPACE

angle. In a celebrated theorem, Gauss demonstrated that a polygon with seventeen sides could be constructed using a straight-edge and compass.

The introduction of a straight-edge and compass does very little to discharge the unease conveyed by the first three of Euclid's axioms, a sense of their uselessness.

Between any two points, it is possible to draw a straight line. This is Euclid straight up, the Euclid of the *Elements*.

Then there is Euclid revised: between any two points, it is possible to draw a straight line using a straight-edge.

Now a reminder: a straight-edge is an edge ending in a straight line. What else could it be?

Whereupon the conclusion that it is possible to draw a straight line with a straight line.

Uh-huh.

IN THE NINETEENTH and twentieth century, mathematicians with briskness and brusqueness in mind, offered Euclid their retrospective assistance in saying what he meant. That business of drawing, producing, and describing? Gone. Euclid's axioms they recast as assertions of existence and uniqueness. There is something, and, by God, there is only one of them.

    1a. Between any two distinct points, there exists a unique straight line.

2a. For any straight line segment, there exists a unique extension.

3a. For every point, there exists a unique circle of fixed radius.

These axioms control the way that the Euclidean universe is filled. They are very powerful: they provide an implicit definition of shape itself. A Euclidean shape is whatever exists by means of Euclid's first three axioms or by repeated application of his first three axioms. The Euclidean constructions were an attempt to capture in physical movement a logical power of the mind. They are gone. It is the arrow of inference that moves. Nothing else.

IN ALL THIS, something is missing, or if not missing, then amiss. Euclid's axioms assume the existence of points. Where else would those straight lines go if not between them? Yet Euclid never once affirms that there exist any points at all, let alone a universe of them.

To Euclid's first three axioms must be added an axiom still more fundamental: that there are points. What is more, there are infinitely many of them, an infinite set of points in modern geometries, a collection or gathering of them, or even a single point allowed Tantric powers of multiplication. Whatever the image, such points exist before anything else does, and in Euclidean geometry, they must exist if anything else does.

A universe of points does not by itself make everything clear where before some things were dark. It is surely false that *any* two points can be joined by a straight line, for unless one thinks of a point as the shrunken head of a straight line, no straight line can join a point to itself. Should one say instead that any two *distinct* points may be joined by a straight line? What makes points distinct? It can be nothing about their internal properties. They have none. To say that two points are distinct only if they are separated in space is to invite the question what separates them? If the answer is a straight line, nothing has been gained.

Euclid's first three axioms lack the sparkle of logical impeccability, spic, but not, perhaps, span. They are doing the work of creation. It would have been a miracle had they done anything more.

EUCLID'S FOURTH AXIOM asserts that

4. All right angles are equal.

This axiom is noticeably different from Euclid's first three axioms. It does not say that anything exists, let alone the right angles. The first three of Euclid's axioms are concerned to get things under way. The fourth is intended to establish a companionable identity among right angles, a brotherhood.

Still, whatever the identity of the right angles, their nature must be encompassed by the first three of Euclid's axioms, together with the decorative ancilla of his definitions.

How might this have worked? Ancient geometers were divided. A right angle?

Geometer A: A right angle is the angle formed when two straight lines are crossed at the perpendicular.

Geometer B: Two straight lines are crossed at the perpendicular when they form two right angles.

Geometer C: Two right angles arise when two straight lines are crossed at the perpendicular.

Geometer D: Gentlemen, gentlemen.

Before right angles are declared equal it would be immensely helpful to know what an angle is in the first place. In this respect, Euclid's axiom is rather like the declaration that all close siblings are competitive. What is a sibling? But then again, what is an angle?

Euclid does say in his eighth definition that "a plane angle is the inclination to one another of two lines in a plane which meet one another and do not lie in a straight line." In his very next definition, Euclid seems to suggest that an angle is what, by his previous definition, an angle *contains*. It is better not to go there. Revising Euclidean geometry early in the twentieth century, David Hilbert considered Euclid's eighth definition and thought that with a bit of polish, it would do nicely, the brass showing through the smudge (see Chapter VIII for Hilbert's system). "Let α be any arbitrary plane," Hilbert writes, "and h, k any two distinct half-rays lying in α and emanating from the point O so as to form a part of two different straight lines. We call the system formed by these two half-rays h, k an angle." An angle is thus a matter of two straight lines suavely exiting a common point.

But Hilbert's definition invites the question when these systems are the same, and when they are different.

Euclid and Hilbert both required some general principle under which angles of any size are judged equal or unequal. A principle is easy enough to contrive. Consider two angles separated widely in space. Two angles and so two systems. Two systems and so four straight lines. Four straight lines and so two points. Two such systems, and so two such angles, are equal if they coincide.

Euclid and Hilbert appear well satisfied.

But to determine whether two angles separated in space coincide, both Euclid and Hilbert must suppose that one of them is moved so that it is imposed on the other. But if moved, then moved in such a way that its *own* angle remains unchanged. This requires a commitment to the homogeneity of space, the idea that as they are moved in space, Euclidean figures do not change in shape. How might this be established without an antecedent account of the identity of their angles?

It is not so much that Euclid's definition is smudged. There seems to be no brass underneath the definition, no matter how much polish is applied to its surface.

THE FIFTH AND final axiom of Euclid's system is more famous than the other four. It is said to have troubled Euclid, who squirmed and turned, wheezed and whistled, before accepting it:

5. If a straight line falling on two straight lines makes the interior angles on the same side less than two right angles, the two straight lines, if produced indefinitely, meet on the side on which are the angles less than two right angles.

The axiom is troubling because it seems to assess the property of parallelism by an appeal to what it is not. The theorem's two straight lines converge at a point; they are not parallel. The subject of Euclid's fifth axiom happens to be lines that *are* parallel. What about them?

An eighteenth-century form of the axiom credited to the Scottish mathematician Francis Playfair is far more intuitive than Euclid's own, and as mathematicians almost at once realized, both versions are logically equivalent:

5a. One and only one straight line may be drawn
     through any point P in the plane parallel to a
     given straight line AB.

The phrase "may be drawn" is permissive when permission is not needed. The axiom affirms that in addition to being unique—"one and only one"—a line parallel to AB and passing through P *exists*.

Point taken, motion adopted. Playfair's axiom says that through a point outside a given line, there exists one and only one line parallel to the given line.

Playfair's axiom completes the axiomatic structure of Euclidean geometry.

It is the last.

EUCLID WAS SAID to be troubled by this axiom because it seemed more complicated than the others. On other accounts, it seemed to him less evident. And still other accounts ascribe to Euclid the contrary conviction that the parallel postulate is simple enough to be a theorem. Those doubts of his are today taken as evidence of Euclid's superb logical intuition. He knew something was wrong, or if not wrong, then not right.

It would be fine to have a Euclidean double willing to join the discussion and enter into the record a few doubts about these doubts. What might he say—this Euclid scrabbling along the path the real Euclid never chose? For one thing, he would, I hope, reject the idea that the fifth axiom is more complicated than the other axioms. Complexity requires a measure or metric, and neither is in the case of these axioms forthcoming. I am sure that a geometer might be found championing the first of Euclid's axioms as more complicated than all the others, just for the heck of it. The fact that Playfair's axiom is simpler than the axiom Euclid introduced is evidence that one and the same axiom may have both a simple and a complicated formulation.

An imaginary Euclid might be equally inclined to disabuse the real Euclid of his *petit soucis* that his fifth axiom might not be evident. *Mais non*! If an axiom is not self-evident, then somehow it must encourage the suspicion, however long deferred, that it might be false. For this reason, no one supposes that the statement that snow is white is self-evident. True, yes; evident on inspection, that, too; self-evident, no. The denial of self-evidence requires some imaginative contingency of the snow is white but it might have been black variety.

But Euclid's parallel postulate is true under the circumstances that Euclid sketched on a brimming dust board, and there is no obvious way in which it might be false. The parallel axiom is obviously not provable; an axiom is an assumption. But neither is the axiom obviously deniable. If it were obviously deniable, it would be possible obviously to deny it.

How would that proceed, that imaginative exercise?

Still, modern mathematicians have seen better and seen further than either of our Euclids. The parallel postulate is anomalous. It is not necessary. It can lapse.

But in every single world in which the parallel postulate fails, it fails either because the underlying space has changed, or because certain common geometrical terms such as distance have been given a new meaning. In the contrived universe that Euclid limned, it does not fail at all.

Chapter VI

THE GREATER EUCLID

> Si les triangles faisaient un dieu, ils lui
> donneraient trois côtés *(If triangles had
> a god, they would give him three sides).*
>
> —Voltaire

EUCLID'S *ELEMENTS* BELONGS to a curious tradition, one that it created and now incarnates—the mountain-climbing pastoral. Mathematicians regard themselves as men of ascent. "Mathematical study and research are very suggestive of mountaineering," the English mathematician Louis Joel Mordell remarked, recalling with satisfaction that when Edward Whymper made the first ascent of the Matterhorn, four of his colleagues perished on the climb. The genre is pastoral because the *Elements* expresses Euclid's intense demand for an idealized world, one in which things are free of friction and inferences smooth as ice. In his influential study, *Some Versions of Pastoral*, William Empson identified the pastoral with the imperative to "put the complex into the simple." What could be more Euclidean? Euclid's *Elements* is that rare thing: its own best example.

If the theorems of Euclid's *Elements* are its peaks, the proofs are a record of his climbs. In some, Euclid gets to the top quickly; in others, he is obliged to grunt and slog, and in these he is like some grizzled old climber recalling how once he was threatened by congelation of the anus. No matter the proofs that he offers, Euclid expects the reader to grasp the drama that they encompass. The proofs communicate tension, release, triumph. They allow the reader to experience the author's discomfort at a distance.

But why do what Euclid has done already? *The base angles of an isosceles triangle are equal.* This is Euclid's fifth proposition. What is the point in proving it all over again? If the question is surprisingly common, the Euclidean answer is uncommonly stern. Euclid regarded demonstrative ascent as its own reward. "There is no royal road to geometry," he contemptuously remarked when some ignominious Ptolemy (Ptolemy Soter) complained that his proofs were too difficult.

No work, nothing gained; no work, nothing learned; no work, nothing.

This, too, is a part of the Euclidean tradition, its moral urgency.

George Mallory attempted to reach the summit of Mount Everest in 1924, and died in the attempt. Under circumstances less demanding than Everest, fellow climbers observed, Mallory would simply swarm up a mountain, like an energetic quadruped. This is not Euclid's way. His proofs are composed of small, mincing, but precise and delicate, logical steps. They must be undertaken one after the other. Not easy. And not easy because the method of proof is one thing; its subject, another. But Euclidean geometry involves the same bifurcation of attention that characterizes the physical sciences. To describe the arc of a cannonball in flight, the physicist, wishing more precision than might be afforded by *there she goes*, must use the analytic apparatus of the differential calculus. The calculus is new; the pattern, old. It is a pattern precisely as old as the *Elements*.

A Euclidean proof does not lend itself to light reading. Each step is easy enough because each step is small enough, but steps cannot be skipped, and retaining in mind all the steps involved in a proof—this is very difficult. A complicated differential diagnosis in medicine, or a brief in contract law, is not easy to read either, but a Euclidean proof, although stripped down because abbreviated by symbols, is difficult in a way in which documents in medicine or the law never are. Pencil and paper are helpful. Diagrams are fine. Euclid's *Elements* is illustrated. Patience is required, no doubt, and beyond that, a taste for alpine altitudes.

Do not take my word for it: consider the master.

PROPOSITION ONE

"On a given finite straight line," Euclid says, it is always possible "to construct an equilateral triangle."

These are the first words of his first proof. In his nineteenth and twentieth definitions, Euclid defines rectilinear figures—figures contained by straight lines—and equilateral triangles: triangles with three equal sides. But neither his definitions, his common notions, nor his axioms say that any of them exist, and in nothing that has come before has Euclid given the slightest indication that it is within his power to make or create them.

In stating his first theorem, Euclid uses the infinitive *to construct*. I have used *to create*. In fact, nothing is either constructed or created. The equilateral triangles are there all along. Euclidean triangles are abstract: the *Elements* does not describe anything physical, and pure Platonic triangles do not come into and out of existence. Euclid's proof reveals them as shapes in the sense that his axioms establish them as objects. The Euclidean maneuver has something in common with revealed checkmates in chess, in which some irrelevant piece is shuffled to reveal the devastating and inexorable combination that it has disguised. And something in common with certain landscape experiences, the hillock having been ascended, the peak long hidden suddenly revealed.

"LET AB BE a given finite straight line."

Using A as a fixed point, Euclid at once deduces the existence of the circle BCD and appeals to his own third axiom to justify this step (Figure VI.1).

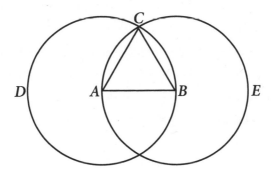

FIGURE VI.1. Proposition one

One circle deserves another: this one, ACE, whose center is B. The third axiom—again.

These circles, Euclid now asserts, must meet at a point C. But by Euclid's first axiom, any two points determine a straight line. Thus the lines CA from C to A, and CB from C to B.

With these straight lines deductively established, the triangle ABC appears; its base is the straight line AB with which Euclid began his proof, and its sides are the straight lines CA and CB. (It hardly matters, I should say, whether the straight line CA is designated as CA or AC—although, I suppose, some logical maniac might ask why it does not matter.)

Now, point A is at the center of the circle BCD. In his fif-teenth, sixteenth, and seventeenth definitions, Euclid has said that given a circle, all of the straight lines from its cen-ter to its circumference are equal. From this, Euclid con-cludes that AC is equal to AB and that BC is equal to BA.

But Euclid has already established that AC is equal to AB. It follows that CA and CB are both equal to AB. Things that are equal to the same thing are equal to one another. This is Euclid's third common notion. It has come in handy, no?

The triangle ACB is thus equilateral.

Done.

SHORT TO BEGIN with, Euclid's proof is psychologically shorter than it appears because it makes only one demand of the reader or the student: that he or she recognize that every radius of a given circle is equal to every other radius. The proof is an elaboration of this idea. Euclid has man-aged perfectly to suggest a powerful machine without do-ing anything more than racing its motor. Even so, it would be unfair to celebrate this little proof without mentioning a few scruples. "It is surprising," the contemporary mathe-matician D. E. Joyce remarked, "that such a short, clear, and understandable proof can be so full of holes."

Holes? And in Euclid too! Witness the very first step that Euclid takes. His axiom establishes that there exists a unique straight line between any two distinct points, but there are no points in the proof to come, or anywhere else in Euclid, to anchor the line AB.

The point of intersection C is the source of a second scruple, for why has Euclid assumed that the straight lines AC and BC must share their vertex at C?

Why, for that matter, has Euclid assumed that three equal straight lines must enclose a triangle? To say that the lines are equal is to say one thing; to say that they are the sides of a triangle is to say quite another.

And scruples about construction—they must be added to what is by now an embarrassing list. In constructing two circles, Euclid has apparently withdrawn the point of his compass from point A and then placed it on point B. Nothing Euclid has said has allowed him to lift his compass at all, if only because no compass is ever mentioned in the *Elements*. But in that case, how has Euclid gotten from A to B?

But enough is enough. Purists, stand down. When all is said and done, Euclid's proof does what a proof must do: it compels belief.

PROPOSITION FIVE

Equilateral triangles are unrelieved in their symmetry. They are the same no matter the angle from which they are viewed. They do nothing and go nowhere. No wonder there are so many of these squat brutes hanging around.

The isosceles triangle is altogether more refined. The sides of an equilateral triangle are all equal. The opposite sides of an isosceles triangle are equal, but each base is on its own. The difference is artistically important. Isosceles triangles have the power to soar. Ecclesiastical architects formed an isosceles triangle with their fingertips and joined thumbs to imagine space tapering upward to the vault of a great cathedral.

There is a connection, Euclid establishes very early in the *Elements*, between the sides and the base angles of an isosceles triangle. That there is *some* connection or other between the sides and the base angles of an isosceles triangle— anyone can see that at a glance. But the connection that Euclid affirms is slightly past any point spontaneously ratified by common sense. If two sides of a triangle are equal, then their base angles are equal too. This is a connection more powerful than some connection or other. The sides of a triangle are straight lines and the angles of a triangle pairs of straight lines. There is a kind of governance at work between them, so that the lines impose their equality on the angles that they subtend. This is not something anyone would see at a glance.

Euclid's fifth proposition is often referred to as the Bridge of Asses. The appeal to a bridge reflects Euclid's illustration (Figure VI.2), which seems to depict a trestle, but those asses have suggested something else, an intellectual bridge that schoolroom donkeys are unable to cross. Yet nothing in Euclid's proof justifies its reputation for difficulty. The proof is neither the simplest possible nor the most elegant, but it does offer an appreciation of Euclid in the fullness of his manner. And something else. Euclid's proof reveals an almost perfect coincidence between the illustration of the theorem and the logical steps needed to demonstrate it. Not all of Euclid's proofs are like this, and not all of his illustrations or diagrams are on their face this revealing.

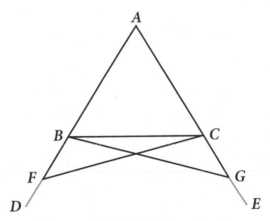

FIGURE VI.2. The Bridge of Asses

"IN ANY ISOSCELES triangle ABC," Euclid affirms, "the angles at the base are equal to one another." What is more, Euclid then adds, "if the equal straight lines be produced further then the angles underneath the base will [also] be equal to one another." The strategy of Euclid's proof is to prove more than what is needed and to take away less than what is proven.

With his original isosceles triangle given in ABC, Euclid constructs a trestle from B to C. Three steps are required for the construction, and some bookkeeping for the argument to come. Euclid first extends the straight lines AB and AC to D and E, respectively. Euclid's second axiom allows straight lines to be extended, and Euclid has just extended two of them.

Permission sought; permission granted.

Euclid then chooses a point F at random on the line BD, and connects the points F and C by the straight line FC. The connection is justified by Euclid's first axiom but the idea of a random choice is an example of Euclid casting out concepts and hoping for the best.

A third step, the last. Choosing a point G on the line AE such that AG is equal to AF, Euclid creates a strut between G and B. His first axiom is in play for the second time. In this, there is nothing strutwise amiss. Yet what justifies the claim that AG is equal to AF? I am glad you asked. Euclid's third proposition affirms that given two unequal straight lines, it is always possible to cut off from the greater a line segment equal to the lesser. AE is the greater line, and AF, the lesser. G is chosen accordingly.

Euclid now requires the assistance of a shady old friend. It is a friend about whom I have already expressed certain reservations, but as Talleyrand said in quite another context, this is no time to be making enemies. Two triangles are congruent, Euclid demonstrated in proposition four, if they coincide in two of their sides and in the angles between them. Using just this proposition, Euclid intends to show that triangles AFC and AGB are congruent. What are friends for?

Already established: AF is equal to AG. There is no need to establish that AB is equal to AC. ABC is, after all, an isosceles triangle. But angle FAG is common to both triangle AFC and triangle AGB. Congruence follows at once, and with congruence, full congress between associated angles. Angle ACF is equal to angle ABG, and angle AFC to AGB.

Again using proposition four, Euclid now shows that BF and FC are equal to CG and GB. But the interior angles BFC and AGB are equal, too. AFC and AGB are congruent, after all. Thus BCF and GBC are congruent as well.

Euclid's conclusion now falls like a ripe peach. Angle FBC is equal to angle GCB, and angle BCF is equal to angle GBC. But the *whole* of angle ABG is equal to the *whole* of angle ACF. Subtracting equals from equals, it follows that the remaining angles ABC and ACB are equal too. These, as Euclid observes with satisfaction, are just the base angles of the isosceles triangle ABC.

Done.

THE GREAT MERIT of Euclid's proof is that it get the job done; its great defect is that it takes so long to do it. The Greek mathematician Pappus provided a proof of the same proposition that is a wonder of efficient elegance. Consider the isosceles triangle ABC and its reflection ACB. The triangle ACB is as close to the triangle ABC as its mirror image. There is in this no surprise. It *is* its mirror image. Whereupon there are a pair of plump reflective identities: AB = AC and AC = AB. Triangle ABC is, after all, isosceles.

And a reminder of what is obvious: angle ABC is equal to angle ACB. It is the same angle.

Therefore, ABC is congruent to ACB, by virtue of Euclid's fourth proposition.

The equality of its base angles follows at once—even faster.

Dulce.

This is lean and cutting as a tooth. But a reservation is in order. Nothing in Euclid's axioms, definitions, or common notions would in the least allow triangles, or any other figures, to be lifted from the plane and reflected back into it.

PROPOSITION FORTY-SEVEN

The Pythagorean theorem affirms that in any right triangle, there is a simple relationship between the length of the triangle's sides a and b, and its hypotenuse h. It is $a^2 + b^2 = h^2$. The sum of the squares of the two sides a and b of any triangle is equal to the square of its hypotenuse. Known apparently to the Babylonians and made known again by Pythagoras in the fifth century BC, the Pythagorean theorem was well known before the ancient world was well worn. On discovering it, Pythagoras on some accounts exclaimed *eureka!* and, on other accounts, sacrificed at once to the Gods. Involving an ox with no interest in geometry, the sacrifice was disgusting, but not so the *eureka*. The theorem's power is obvious. It is grand. In one remarkably limpid statement, it specifies a geometrical relationship among triangles and an arithmetical relationship among numbers. The theorem is powerful enough to compel its converse. Any three numbers a, b, and h such that $a^2 + b^2 = h^2$ determine a right triangle. This is evidence of some haunting unity between the shapes and the numbers, the distinction between the two an accident of appearance, the result of some symmetry broken long ago or an inadvertence in how they are seen.

It is by means of the Pythagorean theorem that the concept of distance comes under general mathematical control. Some control is needed. Distance is an expanse of some sort, an aspect of geometry. It is also a number of some sort, an aspect of arithmetic. It is one or the other or both or neither. But whatever its nature, distance is the answer to the question *how far*, one of the great questions of the human race, inferior only, I suppose, to *how much*. Having been expressed as a conclusion about right triangles, the Pythagorean theorem also describes the distance between any two points in the plane because any two points in the plane can be imagined as a pair of triangular fingertips. This is both magical and marvelous—magical because something answering to the unformed idea of an expanse has been assigned a number, and marvelous because the number has been generated by a short, simple symbolic form. Beyond distance measured in two dimensions, there is, as well, distance drumming its fingertips insistently in three dimensions, and, indeed, in any number of dimensions. The noble family of metric spaces all trace their paternity backwards to the Pythagorean theorem.

The symbolic form $a^2 + b^2 = h^2$ algebraically abbreviates any number of tedious arithmetical facts of the $3^2 + 4^2 = 5^2$ variety. In a, b, and h, it makes use of indeterminate symbols, and in $a^2 + b^2 = h^2$, it imposes three arithmetical operations on the numbers that they designate. It took mathematicians a very long time before they could enter into that dizzying world in which such symbols could be handled with easy confidence. They were handled in the ninth, tenth, and eleventh centuries by the great mathematicians of the far-flung Arabic archipelago, but not with easy confidence, and until quite recently, the handling came hard. Even the greatest of mathematicians had the suspicion that in algebra, as in *The Sorcerer's Apprentice*, symbols they had mastered in one context could prove ungovernable in another.

The apparatus of modern algebra was not available to Euclid. The *Elements* is a treatise of limited symbolic reach. Euclid is content to name points, lines, and various figures in an obvious way, but not once does he step back from the names to consider a more flexible scheme, a more elegant apparatus.

Euclid's proof of the Pythagorean theorem is therefore geometrical. There are no numbers, and no number is squared, but in his geometrical proof, Euclid found a way to convey the arithmetical facts without mentioning them. A great mathematical theorem has many faces. It is one of the ironies of intellectual history that the Pythagorean theorem, which suggests the unity of geometry and arithmetic, should in Euclid's hands receive purely a geometrical proof, almost as if the master could not quite see that in mathematics, as in water, there is never only one side or the other.

IN ANY RIGHT-ANGLED triangles, Euclid says, the square on the side subtending the right angle is equal to the squares on the sides containing the right angle. Two things are equal to one. Not numbers, but shapes. But if shapes, then numbers, too. The area of a square expresses the squaring of a number.

Throughout Euclid's proof, the right triangle ABC exerts a dark magnetic force on his imagination (Figure VI.3). Its sides AB and AC are *the* sides of the Pythagorean theorem, and its hypotenuse BC, *the* hypotenuse. Everything else in the diagram is ancillary and will, when the proof is complete, be allowed to fall away, like the scaffold supporting an arch.

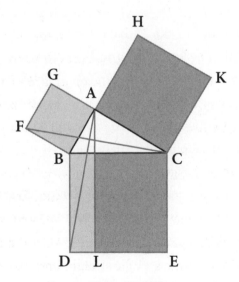

FIGURE VI.3. Euclid's setup

There are three sides to a triangle and three squares in the proof. They are BDEC, which is drawn on the line BC; BFGA, on the line AB; and ACKH, on the line AC. Euclid justifies the construction of these squares by means of his forty-sixth proposition: "On a given straight line, to describe a square." Better because better: On a given straight line, there *exists* a square.

A single triangle has accommodated an entourage of bloated bodyguards. What now; what next?

Proposition forty-seven is the first theorem in which a masterful Euclid imposes on his readers the distinction between strategy and tactics. His tactics involve two sets of congruent triangles. They play the role of proxies. They are congruent, those proxies, and so the same. But as Euclid will show, they are also equal to squares or parts of squares. The strategy of his proof thus involves a feint toward incidental objects, followed by a sweep toward the theorem's essential identities, the axis of his attack curving like a scythe.

IN ORDER TO obtain the first of his proxies, Euclid drops a line from A to L, one parallel to either BD or CE. He then joins the lines AD and FC. BAC and BAG are right angles. CA and AG constitute a single straight line. Euclid's fourth axiom is in play, and so is his tenth definition (which the reader is invited to rescue from its obscurity).

But BA is also a straight line with respect to AH. And what is more, the angles DBC and FBA are equal because they are right angles.

Euclid now adds the angle ABC both to DBC and FBA. It follows that the whole angle DBA is equal to the whole angle FBC. Euclid's second common notion is now in the game and doing useful work at last.

But look: BD is equal to BC. They are sides of the same square. And FB is equal to BA, for the same reason. The triangles ABD and FBC are thus congruent by Euclid's fourth proposition.

Euclid has finished feinting; and the first half of his proof is complete.

EUCLID NOW ESTABLISHES a conclusion about figures that are not triangles at all. The parallelogram BDL, he argues, is twice the triangle ABD. They have the same base BD and are in the parallels BD and AL.

In justifying this assertion—the only recondite assertion in his proof—Euclid appeals to his forty-first theorem: "If a parallelogram have the same base with a triangle, and be in the same parallels, the parallelogram is double of the triangle."[1]

---

1. The translation of this proposition would be easier to understand were the words "in the same parallels" replaced by the words "are lying within the same parallels." The parallelogram and the triangle, in other words, are bounded by the *same* parallel lines.

By the same reasoning, the square GB is double the triangle FBC. They have the same base in FB and are in the same parallel lines FB and GC. The double of equals is equal to one another. It follows that the parallelogram BDL is equal to the square GB.

Euclid now repeats his reasoning. If AE and BK are joined, he asserts, two new congruent triangles appear in HBK and AEC. The parallelogram CL must be equal to square HC.

A farewell to those fabulous feints is in order. They have done their work. Euclid's attack is now direct; it is straightforward; and his proof proceeds by the solid, time-tested tactic of putting two and two together. The square BDEC is equal to its parts in EL and CL. But CL is equal to the square GL; and EL to the square AK. When reassembled, the square BDEC, having been divided for purposes of proof, is equal to GL and AK.

The square on the side BC is equal to the squares on the sides BA and AC.

Done.

See that, Son?

Yes, Sir.

Attaboy.

Chapter VII

VISIBLE AND INVISIBLE PROOF

Reductio ad absurdum, *which Euclid loved so much, is one of a mathematician's finest weapons. It is a far finer gambit than any chess play: a chess player may offer the sacrifice of a pawn or even a piece, but a mathematician offers the game.*

—G. H. HARDY

Some paint comes across directly onto the nervous system and other paint tells you the story in a long diatribe through the brain.

—FRANCIS BACON

IN ANTOINE WATTEAU'S wonderful painting *Jupiter and Antiope*, a tense and muscular Jupiter has withdrawn Antiope's silken robe from her sleeping body, and, of course, bound by durable pigments, he does nothing more, the poor brute forever locked where Watteau left him, lost in longing and fuming with impatience.

This is the great limitation of the Western pictorial tradition. The plane is static. Nothing moves. It is a limitation aching to be violated. In the eighteenth and nineteenth centuries, inventions appeared in which a series of stiff paper scraps would, when rapidly flipped, create a fragile illusion of real life. Adults were enchanted, children amused. There was the zoetrope, the magic lantern, the praxinoscope, the thaumatrope, the phenakistoscope, and the flip book. Praxinoscopic visionaries could see that a series of mounted scraps might one day depict Jupiter in all his massive muscular force doing something more than crouching in impatience. Whatever the unfolding that the cinema reveals, it is one prefigured in the experience that the sophisticated imagination brings to the pictorial plane itself. A great painting invites its own analytical continuation, an arrangement of two-dimensional shapes allowed during a moment of aesthetic fantasy to shed its confinement and enter into the future or the past. In commenting on John Ruskin, the art historian Kenneth Clark appealed to a superiority that allowed him "to conjure images vividly in the mind's eye." Whatever Ruskin's superiority, it embodies a power that in part we all share, the ability in looking at a painting to wriggle out of the present and slip into the stream of time.

The analytic continuation of a great painting very often controls its aesthetic properties and so its natural critical vocabulary. Watteau's *Jupiter and Antiope* is filled with *tension*—the obvious word—and its arrangement of shapes unstable, if for no better reason than the discomfort Jupiter is shortly to feel in his right arm. Johannes Vermeer's *View of Delft* is, by way of contrast, serene. Projected into the future or recovered from the past, it hardly changes, the river passing into an open canal, the clouds, the reflections on the water's surface, the sand—not so much timeless as indifferent, a flow, a ripeness, too.

THE ELEMENTS IS unusual as a mathematical treatise in that it is meticulously illustrated. For every theorem, there is a picture; and with rare exceptions, the pictures are marvels, the *Elements* providing its readers with a series of ingenious geometrical tableaus: triangles, circles, squares, rectangles, lines crossed or in parallel, the stable and familiar shapes of art and architecture, each presented in isolation, a pedagogical handmaiden to the text, the work of a masterful teacher who knew just when the confidence of his students was about to sag. It may be possible to acquire the *Elements* without once attending to its illustrations, but no one has done so.

Like Watteau's *Jupiter and Antiope* or Vermeer's *View of Delft*, there is nothing in the *Elements* that corresponds to what in life is a fluid ever-changing succession—*those images that yet, fresh images beget.* The illustrations are essential because they are a beginning. "Who could dispense with the figure of the triangle, [or] the circle with its centre?" David Hilbert asked in 1900. The axioms have nothing to do with it. "We do not habitually follow the chain of reasoning back to the axioms," Hilbert observed. "On the contrary we apply, especially in first attacking a problem, a rapid, unconscious, not absolutely sure combination, trusting to a certain arithmetical feeling for the behavior of the arithmetical symbols, which we could dispense with as little in arithmetic as with the geometrical imagination in geometry."

It is in the rich and fascinating interplay between the logical structure of his theorems and their brilliantly contrived illustrations that Euclid's art comes most alive.

Yes, *alive*; yes, *art*.

Do we know whether Euclid composed his own illustrations? We do not. The manuscript trail goes cold in the Middle Ages, no more than scraps found earlier, shuffling antiquarians fingering them absently in Cairo or Baghdad and then consigning them to cedar cases. And this is another aspect of the *Elements*, the enigma of the book, the identity of its author.

EUCLID'S TWENTY-SEVENTH proposition affirms that if a straight line EF falling on two straight lines AB and CD makes the alternate angles AEF and EFD equal, then AB is parallel to CD (Figure VII.1).[1] Some stripping of the theorem is required. A straight line EF falls on two straight lines AB and CD. Said once, it need not be said again. The three straight lines are like Somerset Maugham's three fat women of Antibes: they are there. The stripped-down theorem: if AEF is equal to EFD, then AB is parallel to CD.

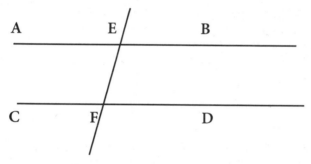

FIGURE VII.1. Proposition twenty-seven

Figure VII.1 tells us no more than this: angles are equal; lines are parallel. It is monotonous. If the figure is understood as a temporal slice—one depicting a figure frozen at one moment—might extending it into the future reveal

---

1. Euclid's twenty-seventh proposition may suggest Euclid's original parallel postulate. This is incorrect. Euclid's twenty-seventh proposition is inverse to Euclid's parallel postulate and so logically equivalent to its *converse*.

something more of its inner tensions, the balance of forces leading to the theorem itself? Crouching in Watteau's oil, Jupiter might moments later be imagined pouncing, but in projecting Figure VII.1 into the next moment, we see nothing that we have not seen before. The thing is as it was. Its angles are equal, its lines parallel.

Just where is that "rapid, unconscious, not entirely sure" visual intuition when it is needed? Needless to say, Figure VII.1 is *not* the diagram Euclid used.

EUCLID'S TWENTY-SEVENTH proposition says that if AEF is equal to EFD, then AB is parallel to CD. What has been stripped down may now be stripped bare: if *P*, then *Q*. It is only when Euclid goes bone deep that he is able to observe the logical space in which his arguments and illustrations fuse completely.

Hypothetical propositions contain two propositions in *P* and *Q*; there is correspondingly a fourfold region of logical space which they may cohabit. *If P, then Q* is front to back, and *if Q, then P* is back to front. One is the converse of the other. A proposition and its converse are logically independent; they are free to go their separate ways. *If ~P, then ~Q* is again front to back (with ~*P* meaning "not *P*" and ~*Q* meaning "not *Q*") and is called the inverse of *if P, then Q*. The converse and inverse of a given proposition are logically equivalent. There is no distance between them;

they say the same thing. And finally, there is *if ~Q, then ~P*, the contrapositive of *if P, then Q*, the coupling of conversion and inversion (rather a desperate description, now that I think about it). A proposition and its contrapositive are logically equivalent.

Although Euclid begins his proof with proposition twenty-seven taken straight-up, his argument depends on its contrapositive: if AB is not parallel to CD, then AEF and EFD are not equal.

In order to demonstrate this proposition, Euclid undertakes a maneuver that is common throughout mathematics and therefore throughout geometry. By dividing his mind, he assigns to one half the position he wishes to rebut, and to the other half, the ensuing right of ridicule. The technique is known as *reductio ad absurdum*, or proof by contradiction. Euclid's strategy is to prove that a proposition is true by assuming that it is false, and then demonstrating what a mess that makes.

From the assumption that the contrapositive to his twenty-seventh proposition is false, Euclid will show that AEF and EFD are equal *and* that they are not.

This is the mess.

Euclid's proof is not self-contained. If it were, there would be no reason to place twenty-six proofs before it. Proposition twenty-seven employs Euclid's sixteenth proposition and his nineteenth and twenty-third definitions. The sixteenth proposition says that in any triangle, if one of the sides is extended, the exterior angles must be greater than either of the interior and opposite angles. Figure VII.2 carries an enviable air of visual authority: angle ACD is greater than angles CBA or BAC. The twenty-third definition has something to say about parallel lines—among other things, if two lines are not parallel, then sooner or later they must meet at a point.[2] His nineteenth definition offers the obvious and expected account of just which figures are triangles.

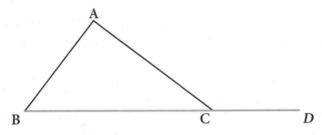

FIGURE VII.2. Proposition sixteen

2. Euclid's twenty-seventh proposition is logically equivalent to Euclid's sixteenth proposition, something that the logician August De Morgan observed in the nineteenth century. Euclid might well have begun with his sixteenth proposition and after demonstrating it, arrived at his twenty-seventh proposition by contraposition. The resulting proof would have been impeccable, but it would not have directly mentioned those parallel lines bound for far places that figure in his twenty-seventh proposition. To get them to come forward, he would have had to reverse logical steps and restore the original proposition.

Euclid is ready now to argue. His claim is that if AB is not parallel to CD, then the angles AEF and EFD shown in Figure VII.1 are not equal. So begin with this. And suppose the proposition false. If false, its antecedent must be true: AB is not parallel to CD. And if false, its consequent must be false: AEF and EFD are equal.

If AB is not parallel to CD, then these lines must meet at a point G. It is here, and only here, that Euclid's own diagram plays a role. Figure VII.3 does one thing. It illustrates the supposition that AB is *not* parallel to CD, the antecedent—and *only* the antecedent—of the proposition Euclid means to reject. The figure is as spare and unforgiving as a sneer. It does not show lines AB and CD converging toward G. Were Euclid to have been interrogated on this point, he might have said with perfect aplomb, Why should I have bothered?

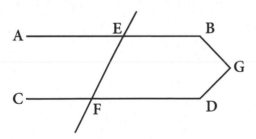

FIGURE VII.3. Proposition sixteen, Euclid's diagram

Why *should* he have bothered?

Given that AB and CD meet at G, Euclid next constructs the triangle GEF. He requires his nineteenth definition and his first axiom, the latter to connect the dots and the former to say what they mean. Euclid's sixteenth proposition affirms that if in any triangle one of the sides is extended, then the exterior angles must be greater than either of the interior and opposite angles. The denial of Euclid's twenty-seventh proposition has encountered Euclid's antecedent proof of proposition sixteen. The turning point of the proof has arrived.

And while Euclid provides no diagrams to illustrate this point, a diagram is easy to contrive (Figure VII.4). The balance of Euclid's proof is now a matter of stating the obvious. From his sixteenth proposition, Euclid has concluded that AEF and EFD *cannot* be equal. But from his assumption that his twenty-seventh proposition is false, Euclid has also concluded that AEF and EFD *must* be equal. The result is the foreseen shambles: AEF and EFD are both equal and unequal.

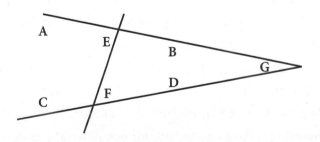

FIGURE VII.4. Contrapositive to proposition twenty-seven

We may allow Euclid to dissolve the distinctions in his mind between back and forth. The denial of proposition twenty-seven had led to a contradiction. Therefore the proposition must be true. Euclid has persuaded himself of himself.

EUCLID'S *ELEMENTS* REMAINS locked in the world before our own, and it thus demands the kind of richly imagined embedding still demanded by Watteau. The painting or the Euclidean diagram is a slice of a temporal continuum, the mind placing in proper perspective the slices that have gone before and those that go afterward.

The purpose of a proof is to compel belief, and to compel his reader's belief, Euclid has invested his twenty-seventh proposition with pictorial life. His diagram and the logical structure of the theorem it expresses undergo fusion. Far more than the axiomatic method alone, it is this fusion that is at the heart of Euclid's method.

In Figure VII.1, lines drawn parallel strike off for the infinite, always together, and by the same distance, too, but always apart, doomed thus to travel in companionable isolation throughout the whole of space. Figure VII.3, by contrast, expresses the contrapositive kernel of Euclid's argument. It offers a lucid and balanced view into the future, but a view narrowed to a single pictorial and geometrical point. The lines are not parallel; therefore, they meet at a point.

Euclid's single diagram leads backward and forward by means of a sequence that Euclid expects the reader to create and then complete, a night series of shapes, fluid and fantastic, the day's unrelenting logic dissolved. Parallel lines bend toward one another, drooping in defiance of the facts, and deflected from parallel position, they converge toward some point G, and converging toward some point G, they register an effect at a distance, and an effect registered, angles change, and the angles changing, "Well, you know or don't you kennet or haven't I told you every telling has a taling and that's the he and the she of it, Look, the dusk is growing. My branches are taking root."[1]

Figures VII.1 and VII.3 depict worlds in counterfactual collision. Lines that are parallel in Figure VII.1 are *not* parallel in Figure VII.3. But worlds in collision on the level of the image represent worlds in collusion on the level of the theorem. Having grasped his proof, you will, Euclid is persuaded, understand its illustration, for with his exquisite power to unify the logical structure of his proof and the diagrams by which the proof is conveyed, he has illustrated a temporal flow.

---

1. James Joyce, *Finnegan's Wake*, Library of the University of Adelaide, South Australia, 2005, p. 213.

As far as Euclid the Magician is concerned, nothing more need be done. He has gotten you to do something rapid, unconscious, and not entirely certain.

As far as Euclid the Logician is concerned, nothing more need be added. He has gotten you to see a sequence of propositions hurtling toward a contradiction.

As far as Euclid the Geometer is concerned, nothing more need be done or added.

THE PROPOSITIONS THAT Euclid demonstrated in the *Elements* ascend by number, and the numbers are a reasonable guide to their difficulty. Euclid's twenty-seventh proposition retains something of the obvious. It encourages the student (or the reader) to a concurrent grunt of affirmation. The theorem is dramatic nevertheless in its reach and power. It draws a connection between a pair of equal angles on the one hand and a pair of parallel lines on the other.

A look is enough to gauge the character of an angle, but no look, however lingering, does much to determine the character of parallel lines. Straight lines are parallel if they never meet. Within the Euclidean plane, *never* goes on and on. How is the geometer to establish that lines that never met will never meet? Once they have passed the last point of public inspection, lines that seem parallel might willfully undertake an unexpected decision to draw close after all.

But equal angles are equal locally, visible in the here and now. By checking the angles made by certain straight lines, the geometer may determine their parallel character once and for all. There is no need to track them to the back of beyond.

This theorem is interesting without in any way being extraordinary. What is extraordinary is what is so often hidden in the *Elements*: the rich ensemble of instruments that Euclid has employed to serve his ends. The proof of proposition twenty-seven is entirely a matter of the scant few lines needed to move with logical assurance from Euclid's premises to his conclusions. But like an army, every one of Euclid's theorems carries a long logistical tail: its apparatus of propositions, axioms, definitions, common notions, and rules of inference. And its illustrations, those diagrams that provide an intuition that is "rapid, unconscious, and not entirely certain."

No part of this immense logistical tail by itself compels belief or elicits assent. It was Euclid's genius to grasp the whole and to trust in the reader to follow what he had grasped.

Chapter VIII

THE DEVIL'S OFFER

> *Algebra is the offer made by the devil to the
> mathematician. The devil says: "I will give
> you this powerful machine, it will answer
> any question you like. All you need to do is
> give me your soul: give up geometry and
> you will have this marvelous machine."*
>
> —MICHAEL ATIYAH

I F THERE ARE numbers *and* there are shapes in nature,
which comes first?

Common sense: neither.

First in *what*?

Although Euclid's *Elements* is a treatise about geometry,
some idea of the natural numbers 1, 2, 3, . . . is present as
background. It could not be otherwise. Euclid talks of triangles, after all, meaning more than *one*, and there is *one*
and only *one* line parallel to a given line specified by Playfair's axiom. The natural numbers are among the common
notions "on which all men base their proofs." The reverse is

true as well. No mathematician could study arithmetic if the numerals did not have stable geometrical properties. Imagine trying to prove that 2 plus 2 is 4 and seeing the numeral "2" undertake a sinuous deformation on the blackboard, or in the mathematician's mind, vanishing, perhaps, at the very moment of intellectual triumph.

The oil of compromise having been spread, the question, of course, remains. Which *does* comes first, geometry or arithmetic—first in the sense of being more fundamental, as bread is more fundamental than butter, and thus first in the sense that geometry may be derived from arithmetic, or arithmetic from geometry?

THE NATURAL NUMBERS: 1, 2, 3, . . . Although the number 1 is smaller than all the rest, there is no number greater than any of the others. If there were such a number n, the number $n + 1$ would be greater still. Does it follow that the natural numbers are infinite? The great Gauss offered a warning. "I protest," Gauss remarks, "against the use of infinite magnitude as something completed, which is never permissible in mathematics. Infinity is merely a way of speaking." The proper way to speak is to speak guardedly. The natural numbers are *potentially* infinite. The mathematician ascends from 1 to 2, from 2 to 3, and from n to $n + 1$. But unless counting can go on *forever*, this analysis can hardly do justice to the natural numbers; and if it can go on forever, why not admit the infinite once and for all and be done with it? Georg Cantor, the creator of set theory, argued in the late nineteenth century that the set of natural numbers comprises something infinite all at once, a great thing, complete in its luxuriance. What Cantor could not say is just how the human mind gains access to the infinite if not by climbing the staircase of the numbers, one step at a time. The natural numbers comprise an infinite set. Point to Cantor. Access to the infinite is incremental. Point to Gauss.

This is not Euclid's way.

In Books V, VII, and X of the *Elements*, Euclid talks about the numbers. "A unit," he says, "is that by virtue of which each of the things that exist is called one." A number is a "multitude composed of units." Ancient commentators, writing before and after Euclid, suggested that a unit was the least or smallest answer to the question *how many?* The answer: just *one.* They were well aware, of course, that whatever *one* might be, it, too, could be divided, but this, they argued, would only return the mathematician to whatever multiplicities the number one was supposed to resolve in favor of the least of them.

Although Book VII of the *Elements* addresses the numbers explicitly, the book is logically subordinate to Book V. The numbers of Book VII are, in fact, the magnitudes of Book V. Book X of the *Elements* is in an act of double deference subordinate to both. A rare lapse in organization is present throughout. Book V introduces Euclid's theory of magnitudes. It is widely considered one of Euclid's masterpieces, although whether by Euclid or whether a masterpiece are other questions. Historians of Greek mathematics now suggest that Euclid's theory is due largely to the Greek mathematician Eudoxus of Cnidus. Having anticipated the calculus with his method of exhaustion, Eudoxus also appears to have anticipated the real numbers with his theory of proportions. In his treatise *Elementary Geometry from an Advanced Standpoint*, the mathematician Edwin Moise argued that with respect to the real numbers, modern mathematicians had no need to create what Eudoxus had known all along.

With a unit fixed—a unit *chosen*—Euclid gains access to a geometrical simulacrum of the natural numbers and to the rational numbers or fractions, such as ½ or ⅓. The number 7 corresponds to 7 units laid tail to head, and the number ⅐, to the ratio of 1 unit to 7 of them.

In the progression of the *Elements*, a magnitude represents a new idea, but the easy familiarity with which Euclid handles it suggests that he thought of magnitudes as natural aspects of his system—friends of the family. Euclid never quite says what a magnitude *is*, but the general idea is of extent, an occupation of space, an expanse. Whatever the expanse encompassed by a unit, it corresponds in the plane to a straight line fixed between two points. This is how Euclid illustrates every one of his arithmetic ideas. It follows that the natural numbers, since they are without end, must correspond to the indefinite production of a straight line.

"It seems that the aging Plato," René Thom remarked (with some indifference to the English language), "considered this type of generativity to be of the type [of] discrete generativity [that is characteristic] of the sequence of the natural integers." The very old Plato may have murmured a word into the very young Euclid's ear; if not, his words were still within hearing distance.

A striking simile is at work in the *Elements*, one that is today anachronistic. Euclid had found the source of arithmetical generation in a geometrical object. The shadows play on the Euclidean plane. Beyond, there is the play of real things in the real world. Whence the simile. The Euclidean line moves in Euclidean space as a physical object moves in physical space. The idea occurred to many Greek mathematicians. It is the basis for the Greek scheme of geometrical algebra. As it works its way through Euclid's system, it introduces a degree of contrivance into Euclid's thoughts. The Euclidean line flows through the points that it touches, but it can be divided into discrete segments only by means of the geometer's retrospective artifice. From the outset, the Euclidean idea of number conflicts with the simile by which it is explained. In the end, what may seem nothing more than a conceptual conflict grew until it threatened the integrity of the Euclidean system itself.

IT IS NOT disrespectful—is it?—to say that geometrical algebra has in Euclid's hands all of the elegance of bears chained and taught to dance. In his proof of the Pythagorean theorem, Euclid ignores the algebraic equation in which the facts are so easily expressed—$a^2 + b^2 = h^2$—and occupies himself with the construction of those rather oafish squares, seeing in their area the secret to the theorem's meaning. It is a clumsy business. The first proposition of Book II of the *Elements* affirms that the area of a given rectangle is equal to the sum of its subrectangles. This is in algebraic terms, the distributive law $a(b + c + d) = ab + ac + ad$, where a, b, c, and d are numbers. The rectangles are illustrations; they get in the way. Euclid takes geometrical algebra as far as he can go, but by the time he gets to where he is going, the tide must already have begun to turn. And while it took a long time to flow out, in the end it flew out, until mathematicians universally acknowledged the imperatives of analytic geometry, the countercurrent.

Writing in the seventeenth century, René Descartes created analytic geometry in a work titled *La Géométrie*. Descartes was not quite sure what he was doing. His great work he left almost as an afterthought. In analytic geometry, the Euclidean plane is made accessible, and so it is opened up, by means of a coordinate system. A point is chosen arbitrarily, the origin. Since all points are in the end the same, it hardly matters *which* point is chosen. Whatever the point, it corresponds to the number zero. Thereafter, the point is bisected by two straight and perpendicular lines, the coordinate axes of the system. The positive natural numbers run from the origin out to infinity, the negative numbers run out the other way, back-street boys to the end, and precisely the same scheme is repeated for the vertical axis, making four line segments starting at zero and proceeding inexorably to the edge of the blackboard and the space beyond.

Any point in the plane may now be identified by a pair of numbers (Figure VIII.1). Hidden previously in the sameness

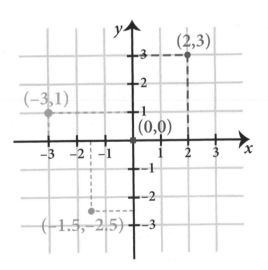

FIGURE VIII.1. Euclidean coordinate system

of space, a point acquires a vivid arithmetical identity. It is *the* point corresponding to two numbers, where before it was some drab or other, anonymous. Once points have acquired their numerical identity, the mathematician can deploy the magnificent machinery of algebraic analysis to endow Euclidean geometry with a second and incomparably more vivid form of life.

ARITHMETIC IS THE place where the numbers are found, and algebra, the place where they are treated in their most general aspects. The points and straight lines of Euclidean geometry—make of them what you will. They are undefined. Now that a geometrical point has been identified with a pair of numbers, a straight line can be defined by the equation $Ax + By + C = 0$, where A, B, and C are numerical parameters, fixed place markers, and x and y variables denoting points resident on the line.

Did Euclid have circles to command? He did. A circle whose center is at the point (a, b), and whose radius is R, is perfectly and completely described by the formula $(x-a)^2 + (y-b)^2 = R^2$. The identity of the circle has been dominated by a numerical regime: its center is a pair of numbers; its radius, another number, and its circumference, an endless succession of numbers.

Analytic geometry has the power to depict a great many familiar geometrical shapes, such as the parabola, the ellipse, and the hyperbola. There is also the *cardioid*, its penciled heart emerging from the billet-doux of $(x^2 + y^2 + ax)^2 = a^2 (x^2 + y^2)$ (Figure VIII.2).

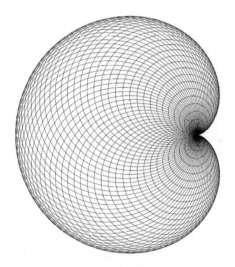

FIGURE **VIII.2.** The cardioid

There are curves that look like a woman's smile, or the valley between hills, or the exuberant petals of some tropical flower.

There is an abundance.

IN A LITTLE book titled *The Coordinate Method*, a troika of Russian mathematicians (I. M. Gelfand, E. G. Glagoleva, and A. A. Kirillov) offers this account of analytic geometry: "By introducing coordinates, we establish a correspondence between numbers and points of a straight line." They then add: "In doing so, we exploit the following remarkable *fact*: There is a unique number corresponding to each point of the line and a unique point of the line corresponding to each number" (emphasis added). The remarkable fact to which they appeal is often described as the Cantor-Dedekind axiom, although how a fact could be an axiom, they do not say.

It hardly matters. There is no such fact, and neither is there any such axiom. There are some numbers that no accessible magnitude can express. In Book X of the *Elements*, Euclid offers a proof that this is so, one based on an earlier proof attributed to the Pythagorean school. He fails only to notice that what he has done constitutes an act of immolation.

The hypotenuse of a right triangle whose two sides are both 1 is, by the Pythagorean theorem, the square root of 2. The square root of 2 is neither a natural number nor the ratio of natural numbers. A proof simpler than Euclid's own proceeds by contradiction. Suppose that the square root of 2 *could* be represented as the ratio of two integers so that $\sqrt{2} = a/b$. Squaring both sides of this little equation: $2 = a^2/b^2$. Cross-cutting: $a^2 = 2b^2$.

Now the fundamental theorem of arithmetic affirms that every positive integer can be represented as a unique product of positive prime numbers. A prime number is a number divisible only by itself and the number 1. Known to the Greeks, this theorem was known to Euclid. It was widely known; it had gotten around.

The little equation $a^2 = 2b^2$ is shortly to undergo a bad accident. Whatever the number of prime factors in a, there must be an *even* number of them in a^2. There are twice as many. Ditto for b^2. But the number $2b^2$ has an *odd* number of prime factors. The number 2 is, after all, prime. Either the square root of 2 is not a number, or some numbers cannot be expressed as natural numbers or as the ratio of natural numbers.

This is the bad accident.

The consequences are obvious. If the Euclidean line does not contain a point corresponding to the square root of 2, how can the Cantor-Dedekind axiom be true, and if it does, how can the line be Euclidean?

THE NATURAL NUMBERS 1, 2, 3, . . . constitute the smallest set of numbers whose existence cannot be denied without commonly being thought insane. Whatever their nature, nineteenth-century mathematicians discovered how numbers beyond them might be defined and so made useful. A single analytical tool is at work. New numbers arise as they are needed to solve equations that cannot be solved using numbers that are old. Zero is the number that results when any positive number is taken from itself and so appears as the solution to equations of the form $x - x = z$. The negative numbers provide solutions to equations of the form $x-y = z$, where y is greater than x. The common fractions, numerator riding shotgun on top, denominator ridden below, are solutions in style to any equation of the form $x \div y = z$.

There remained equations such as $x^2 = 2$. The equation is there in plain sight. What, then, is x? The answer proved difficult to contrive. The Greeks endeavored to find a sense suitable to the square root of 2, but they did not entirely succeed, and beyond the imperative of solving this equation, mathematicians had no common currency with which they could easily pay for its solution. They had nothing in experience.

In the late nineteenth century, Richard Dedekind defined the irrational numbers—how did numbers that are not rational come to be *irrational?*—in terms of *cuts*, a partitioning of the integers into two classes, A and B. Every number in A, Dedekind affirmed, is less than any number in B, and what is more, there is no greatest number in A. The cuts themselves he counted as new numbers, the enigmatic square root of 2 corresponding to the cuts A and B in which all numbers less than the square root of 2 are in A, and all those greater than the square root of 2 are in B. Dedekind's cuts are not the sort of animals one is apt to find in an ordinary zoo. Dedekind's cuts are, it must be admitted, transgendered, their identity as numbers at odds with their appearance as classes. Dedekind demonstrated, nevertheless, that they were what they did not seem to be, and that is number-like in nature. They could be added and multiplied together; they could be divided and subtracted from one another. They took a lot of abuse. They were fine. They were, in any event, more appealing than the supposition that where there really should be a number answering to the square root of 2, there was no number at all.

The formal introduction of the real numbers in the nineteenth century brought to a close an arithmetical saga, one in which numbers that had once inspired unease acquired their own, their sovereign, identity. The positive integers, zero, the negative integers, the fractions, and the real numbers were all in place. They had acquired an indubitable existence in the minds of mathematicians. The system had a kind of abstract integrity. It held together under scrutiny. It was not adventitious.

THE REAL NUMBER system represented the confluence of two triumphs: the triumph of arithmetic and the triumph of algebra. The triumph of arithmetic is obvious. The real number system is a system of real *numbers*. The triumph of algebra, less so. The real numbers satisfy the axioms for an identifiable algebraic structure, what mathematicians call a *field*. The great achievement of nineteenth- and early twentieth-century mathematics was by a python-like compression of concepts to detach the structure from its examples. Writing in 1910, the German mathematician Ernst

Steinitz proposed to make use of fields in an *"abstrakten und algmeinen Weise"*—in an abstract and general sense. A field, he wrote, is a system of elements with two operations: addition and multiplication. That is all that it is. Steinitz then introduced the distinctively new, entirely modern note, the one that marks a decisive promotion of an interesting idea into an independent idea. Never mind the question, the field of *what*? The abstract concept of a field is itself at the *mittelpunkt* of his interests. The examples dwindle away and disappear. The field remains. It becomes itself.[1]

THE AXIOMS FOR a field bind its various far-flung properties together.[2] Their exposition calls to mind the lawyers in *Bleak House* rising to make a point.

—A field is a set of elements, *M'Lud* . . .
—Elements, *M'lud*, anything really.
—Feel it my duty to add, *M'lud*, that there are two operations on these elements . . .
—Beg pardon? Any two distinct operations, *M'lud*.

---

1. Ernst Steinitz, "Algebraische Theorie der Körper," *Crelles Journal* (1910) (my own translation).
2. For which, see my own *One, Two, Three* (New York: Pantheon Books, 2011).

—Feel it my duty to add that there is 0 somewhere, *M'lud*. Yes, here it is.

—Do? It does nothing *M'lud*: $a + 0$ is always a.

—There is a 1, too, *M'lud*. Yes, I have it here. Beg pardon? Nothing. It does nothing *M'lud*: $1a$ is always a.

—Feel it my duty to add a word about inverses, *M'lud*. I have them here.

—Beg pardon? Do? They invert, *M'lud*. Any element plus its inverse is 0, and any element times its inverse is 1.

There is no need to pursue this particular courtroom drama beyond the judge's demand that his attorneys sit down. A field is an abstract object, and so above it all. Still, it is an abstract object whose most compelling example is the ordinary real numbers. An associative law holds force: $a + (b + c) = (a + b) + c$. And so does a distributive law: $a(b + c) = ab + ac$. Identities in 0 and 1, and inverses in the negative numbers and fractions, make possible the recovery of subtraction and division. It is, as lawyers say, familiar fare. A last lawyer rises to remind the judge that the real numbers are ordered. It is always one number before the other, or after the other; it is always, as the judge mutters, one thing or another.

No matter the lawyers, this idea has been a triumph, the second, after the definition of the real numbers themselves. This prompts the obvious question: a triumph over what?

IN 1899, DAVID Hilbert published a slender treatise titled *Grundlagen der Geometrie* (The foundations of geometry). Having for many years lost himself in abstractions, a great mathematician had chosen to revisit his roots. Over the next thirty years, Hilbert would revise his book, changing its emphasis slightly, fiddling, never perfectly satisfied. The *Grundlagen*—the German word has an earthiness lacking in English—is a moving book, at once a gesture of historical respect and an achievement in self-consciousness. In writing about Euclidean geometry, Hilbert was sensitive to the anxieties running through nineteenth-century thought. Well hidden beneath the exuberant development of various non-Euclidean geometries, the anxieties could often seem arcane. But what mathematicians had suppressed was a concern, sometimes amounting to a doubt, that in geometry, the monumental aspect of Euclid's system might all along have disguised the fact that none of it made any sense.

Were the axioms of Euclidean geometry consistent? Or was there buried in the dark flood of their consequences propositions that together with their negations could *both* be demonstrated? To imagine that Euclidean geometry might be *inconsistent* would be to place in doubt more than an axiomatic system, but the way of life that it engendered. Hilbert's *Grundlagen* did not answer this question completely because it cannot be completely answered. Hilbert showed that geometry is consistent *if* arithmetic is consistent, an achievement a little like demonstrating that one building is tall if another is taller, but an achievement nonetheless.

Hilbert undertook the reformation of Euclidean geometry by expanding to twenty Euclid's original list of five axioms. In a remark of some cheekiness, Thom described Hilbert's system as a work of "tedious complexity." The details *are* onerous. Hilbert had found and then corrected a number of logical lapses in Euclid; he was fastidious. Hilbert accepted, as Euclid did, points, lines, and planes as fundamental, bringing them explicitly into existence by assumption. He had already outlined his method in an essay titled *"Uber den Zahlbegriff"* (On the concept of number): "One begins by assuming the existence of all elements (that is one assumes at the beginning three different systems of things: points, lines and planes) and one puts these elements into certain relations to one another by means of certain axioms, in particular the axioms of connection, order, congruence and continuity."

Having set out twenty axioms, Hilbert then steps back to cast a cool, appraising eye on what he has done. There is a change in emphasis, a heightened sense of explicitness. Euclid's analysis is directed toward the world of shapes, but Hilbert has begun to think about the analysis itself, his patient, as so often happens, left droning on the leather couch.

The subtle distinctions needed to make these issues immediate did not exist at the beginning of the twentieth century. Logicians required time to develop them. Hilbert was careful; he made no mistakes in his treatise, but he was not up to date.

A *theory*, logicians now say, consists of a set of axioms together with its logical consequences. Euclidean geometry is a theory, the first in human history. A *model* of a theory consists of the structures in which it is satisfied, a mathematical world, a place in which a theory is at home. Euclidean geometry is satisfied in the Euclidean plane. The simple idea in which theories are juxtaposed to their models makes it possible to ask what models make theories true and whether one theory could be expressed within the alembic of another. It is this idea of re-expression or reinterpretation that Hilbert advanced in his treatise, the tool that he developed.

HILBERT'S *GRUNDLAGEN* IS a work with divided purposes. It is, among other things, a defense of classical analytical geometry.

In thinking about the numbers, Hilbert considered two axioms, the so-called Archimedean axiom, and the so-called completeness axiom. No two axioms have ever been so many times so-called. The first axiom may be found in Eudoxus; it is an implicit aspect of his theory of proportions. The axiom has a simple, powerful intuitive meaning.[2] When it comes to certain numbers, there is no greatest among them and no least among them either. The axiom is satisfied by the rational numbers. The axiom went far in the ancient world, but it did not go far enough. It did not suffice to characterize the real numbers, and for this, the completeness axiom was required.

"To a system of points, straight lines, and planes," Hilbert wrote, "it is impossible to add other elements in such a manner that the system thus generalized shall form a new geometry obeying all of the five groups of axioms." There are as many points on the line as there are real numbers. There are enough to go around. This is not the Cantor-Dedekind axiom, which speaks to a *correspondence* between points and numbers. The completeness axiom is of the enforcement variety. It establishes the existence of those points. It brings them about. It guarantees them. The guarantee makes possible, if not plausible, the techniques of analytic geometry.

But Hilbert's completeness axiom is not an axiom of geometry. The objects that the axiom introduces to complete the points on the Euclidean line are not Euclidean: they are not geometric. They belong to arithmetic and they come from afar.

---

2. I discuss that simple, powerful intuitive meaning in David Berlinski, *A Tour of the Calculus* (New York: Pantheon, 1995).

Having offered an aggressive defense of analytical geometry—here they are, the real numbers, take them or leave them—Hilbert at once revised his tone and tome in order to argue peacefully in favor of a version of Euclidean geometry requiring no direct concourse whatsoever with the arithmetical side of things.

In Books V, VII, and X of the *Elements*, Euclid attempted to see in the lights and shadows of a purely geometrical world the stable figures of arithmetic. He looked very hard, but what he discerned, he never discerned completely. Too many lights and, of course, too many shadows. In the *Grundlagen*, Hilbert justified Euclid and made him whole, *son frère du silence éternel*.

The device that Hilbert employed, he called the calculus of segments. Segments are line segments, as Euclid had supposed. There is no form of arithmetical generation at work. Geometry is paramount. With the kind of patience that prompted Thom to complain that he was bored, Hilbert endowed Euclid's line segments with arithmetic powers all their own. They could be added together, subtracted from one another, multiplied and divided, arranged in continuous proportional arrays, and they could encompass some square roots.

So it could be carried out—*yes*, the old haunting, incomplete, Euclidean arithmetic scheme. It could be carried out, as Hilbert understood perfectly, but not carried out completely. Line segments are not numbers. They may be used to illustrate numbers or to form a picture by which the numbers are understood, but *they* are not *numbers*.

THE FRENCH VERB *engloutir* means to swallow something without chewing—swallowing it whole, an annihilation. Having given up brawling in favor of good works, still another Hilbert is prepared to give up good works in favor of the revolution. This Hilbert is prepared to show that Euclidean geometry may be swallowed by the field of the real numbers. His work complete, Hilbert the Red demonstrated that the points, lines, and planes of Euclidean geometry are actors in an algebraic world not of their making and indifferent to their nature, their impression to the contrary entirely a matter of false consciousness.

THE GEOMETRICAL THEORY that Hilbert presented in the *Grundlagen* is very much Euclid's theory. Hilbert's axioms are more precisely expressed. There are many more of them than may be found in Euclid. The spirit is the same. Euclid had, of course, intended his theory to be interpreted in the Euclidean plane, geometrical axioms satisfied in a purely geometrical model. *This* will not do because *that* did not serve Hilbert's radical agenda. For his own model, Hilbert chose an arithmetic object, one composed of a set of numbers Ω. These numbers begin with 1, and they include all the numbers that may be made from 1 by the operations of addition, subtraction, multiplication, and division, and the numbers $\sqrt{(1 + \Omega^2)}$. This structure generates the real numbers, but it does not generate *all* of them. It is limited. This spare structure also satisfies the axioms for a field. Hilbert made his choice of Ω for the sake of convenience; just a few pages later, he recognized that having introduced some of the real numbers but not all of them, he might well have introduced all of the real numbers and not just some of them.

What follows is swallowing in steps. "Let us regard," Hilbert writes, "a pair of numbers (x, y) . . . as defining a point." What is undefined in geometry has just met what is perfectly defined in arithmetic. In the ensuing friction, any sense of a correlation between points and numbers is lost. There is an *identity*: a point is a pair of numbers. The Euclidean plane gives way. All is dark. The mathematician's attention has swept out from one theory in geometry to another theory in arithmetic.

Euclidean straight lines follow Euclidean points into the void, only to emerge again, reborn in arithmetic. A straight line, Hilbert writes, is a ratio of three *numbers* $(a: b: c)$. An insistent arithmetical identity now imposes itself on the old-fashioned and rapidly receding Euclidean shape. A straight line *is* the ratio of three numbers. The choice of three numbers suggests, of course, the equation for the straight line, the symbols $Ax + By + C = 0$, bringing all of them under the control of a single symbolic form. The numbers are expressed as a ratio because two equations $a_0x + b_0y + c_0 = 0$ and $a_1x + b_1y + c_1 = 0$ define—they *are*—one and the same straight line if the numbers a, b, and c are proportional. In this way, Hilbert has offered an interpretation of the undefined terms of Euclidean geometry in terms of the defined elements of a real ordered field. He has gotten one theory to speak in another theory's voice.

It is like hearing a cat bark.

THE EUCLIDEAN POINT has vanished in favor of pairs of numbers; the Euclidean straight line in favor of triplets of numbers—their ratios. Hilbert is now free to provide an interpretation of Euclid's axioms in arithmetic. Following Euclid, Hilbert affirms that two distinct points A and B always determine a straight line a. This axiom is assumed in geometry, but in arithmetic, it is not assumed at all. It is demonstrable. Hilbert had already gained the power to say what it means for a point to lie on a straight line without ever mentioning points or straight lines at all. "The equation $ax + by + c = 0$," he writes, "expresses the condition that the point (x, y) lies on the straight line $(a : b : c)$."

Euclid's first axiom talks of two points A and B, and one straight line a. The point A is equal to a pair of numbers

(x_1, y_1). Ditto the point B, like A also equal to a pair of numbers (x_2, y_2). The straight line a is equal to a ratio of numbers (a, b, c).

Euclid's first axiom is true in its arithmetical model, Ω, if some equation may be found satisfying both A and B. And, of course, there is. The point (x_1, y_1) lies on the straight line $ax_1 + by_1 + c = 0$. The point (x_2, y_2) lies on the straight line $ax_2 + by_2 + c = 0$. Subtracting the second equation from the first yields $a(x_2 - x_1) + b(y_2 - y_1) = 0$. The parameter c has vanished; a and b may now be banished in the equation $(y_1 - y_2)x + (x_2 - x_1)y + (x_1y_2 - x_2y_1) = 0$. Two points: but one straight line.

What has come forward has come back. The circle is closed. Step by patient step, Hilbert shows how every one of the axioms of Euclidean geometry can be interpreted within a purely arithmetic model. But Hilbert, of course, does more. The fact that every two points determine a straight line is not only true in the real ordered field, it is demonstrable. Geometrical axioms have become arithmetical theorems. By this maneuver, Euclidean geometry has been swallowed by arithmetic, the swallowing giving rise to what are today called Euclidean vector spaces, new structures, ubiquitous throughout mathematics, their sleek compact lines concealing all traces of the annihilation by which they were created.

There is no coordination, no counterpart, no mapping, no scheme of correlation between points and pairs of number. This is because under Hilbert's analysis, there are *no* Euclidean points of old.

"Analytical geometry," the French mathematician Jean Dieudonné observed happily in remarks now famous, "has never existed. There are only people who do linear geometry badly, by taking coordinates, and they call this analytical geometry. Out with them!"

"Down with Euclid," he added, just to be sure.

Chapter IX

THE EUCLIDEAN JOINT STOCK COMPANY

> *A tradition is kept alive only by something being added to it.*
>
> —HENRY JAMES

A T THE BEGINNING of the nineteenth century, the Euclidean Joint Stock Company was wholly owned by Euclid and his *Elements*. If there was among mathematicians a residual uneasiness about Euclid's parallel postulate, it did little to diminish their sense that Euclidean geometry had, over the course of 2,300 years, been valued at its true price. The philosophers agreed. Euclidean geometry, Immanuel Kant argued in *The Critique of Pure Reason*, was not only true—*of course it was*—but necessarily true, an aspect of the human mind, the expression of the way the mind confronted the sensuous world of shapes.

Within sixty years, the famous old company had undergone a dilution of ownership. Kant had doubled his investment at the very moment his investment had been halved in value. And not only Kant. The philosophers had all missed something; theirs was an insufficiency of daring.

By the end of the nineteenth century, it had become clear that there were geometrical schemes in which Euclid's parallel postulate might be replaced by its denial. The long-expected shambles did not emerge. When in 1915, Albert Einstein published the field equations for his theory of general relativity, non-Euclidean geometries acquired a dignified physical standing. Mathematicians such as Carl Friedrich Gauss, Nicolai Lobachevsky, János Bolyai, and Bernhard Riemann, the men who had *made* non-Euclidean geometry, demanded substantial ownership positions in the Euclidean company, and they received them.

EUCLID'S FIFTH AXIOM is one of the few statements in mathematics to have achieved a stable sort of notoriety. It is shady; this everyone understands. And controversial. This, too, is understood. Judging from the popular literature, no one is quite sure why.

Euclid's parallel postulate has today been enveloped by the cloak of its flamboyant history. Euclid made no use of his postulate until he came to his twenty-ninth proposition. It would seem that he had brooded suspiciously. He had put things off. Common sense might suggest that Euclid did not use the parallel postulate before he did, because he had not required it before he had. This is a prosaic view and for this reason not widely entertained.

Ancient commentators wondered whether the parallel postulate might be deduced from Euclid's other four axioms. They sensed its anomalous character. It made them uneasy. They could not entirely say why, and sometimes they were betrayed by their scruples. Proclus rejected a fallacious proof of the parallel postulate by Ptolemy and at once advanced a fallacious proof of his own. Ancient mathematicians often assumed what they intended to show, a circumstance that enlarged their frustrations without ever resolving their anxieties.

Mathematicians uninterested in proving the parallel postulate were often interested in demonstrating that it was equivalent to something else, perhaps in the expectation that swapping things around would reveal something simpler and more compelling. The parallel postulate *is* a proposition too powerful to be exhausted by a single identity. The Pythagorean theorem and Euclid's parallel postulate are the same. The first leads logically to the second; the second leads logically to the first. In Book I of the *Elements*, Euclid demonstrated that $A + B + C = \pi$, where A, B, and C are the interior angles of any triangle. It is his thirty-second proposition. But $A + B + C = \pi$ is logically equivalent to the parallel postulate. Euclid had demonstrated what he had already assumed, the parallel postulate exerting a deforming force on the very structure of Euclidean deduction.

Attempts to prove the parallel postulate continued sporadically over the next two thousand years. Some mathematicians gave the matter a look and, after a few desultory attempts, turned away. The parallel postulate seemed a hard little knot, a twisted root. Mathematicians of the long and brilliant Arab renaissance were as intrigued by the parallel postulate as the Greeks before them had been. Writing in the tenth century, Ibn al Haytham thought that the postulate might require an indirect proof. Euclid had often demonstrated theorems in the *Elements* by assuming that they were false and searching for the contradiction that ensued. Ibn al Haytham did the same; he searched for the shambles. He found nothing, his assumption, for the sake of argument, that the parallel postulate was false leaving everything the same, glassy and undisturbed.

Omar Khayyám, the author of the *Rubiyat* and a mathematician of distinction, made no effort to prove the parallel postulate. He was among the mathematical swappers. And he found nothing of interest:

For "Is" and "Is-not" though with Rule and Line
And "Up" and "Down" by Logic I define,

Of all that one should care to fathom,
Was never deep in anything but—Wine.[1]

TURNING HIS HAND to geometry after the rigors of his Je-
suitical education, the seventeenth-century Italian mathematician Girolomo Saccheri published a treatise in 1733
optimistically titled *Euclides ab omni naevo vindicatus* (Euclid vindicated of every flaw). Very properly, he dismissed
efforts to derive Euclid's parallel postulate directly from Euclid's other axioms. His Jesuit education allowed him to recognize that a dead end was dead. He turned the other way,
assuming the parallel postulate false in order to derive that
long fabled, fatal contradiction from his assumption. Saccheri was able to demonstrate many interesting propositions. He came very close to discovering non-Euclidean
geometry; his Italian admirers, such as Beltrami, gave him
as much credit for failing as they would have given a French
mathematician for succeeding. Saccheri was shrewd, able,
and penetrating. The contradiction that he anticipated, he
never found.

It did not occur to him that it was not there. It did not
occur to *anyone*.

---

1. Omar Khayyám, *The Rubáiyat of Omar Khayyám*, translated by
Edward FitzGerald (San Francisco: W. Doxey, 1898).

There things stood until Gauss began to wonder, in the last years of the eighteenth century, whether the very attempt to derive Euclid's parallel postulate from Euclid's other axioms, or from anything at all, might not have been doomed, not simply difficult. No derivation had been achieved, Gauss reasoned, because none was possible. He was yet an adolescent, a young man of sixteen or seventeen. Remarkable ideas came to him often. Gauss declined to publish his thoughts later in his life, preferring, when others had published theirs, to remark that he had known it all along.

And often he had.

It is possible, of course, to place Euclid's parallel postulate in protective custody, the little lunatic locked in a padded cell, so that the theorems that result are derived entirely from the first four of Euclid's assumptions. The system is called neutral geometry. It is hardly a description engendering excitement. Still, Euclid had demonstrated his first twenty-eight propositions by means of the axioms of neutral geometry; if nothing else, neutral geometry does the obvious. But it is the parallel postulate that is all mad glitter and glow and it does the rest.

The nineteenth-century mathematicians who had come to suspect the parallel postulate could choose whether to embrace it as an axiom or discard it as an impediment. Euclid had embraced his own. Bolyai and Lobachevsky rejected what Euclid had embraced, arguing for an unembarrassed denial of the parallel postulate. The possibilities that emerged were unsettling. Over the course of two thousand years, Euclid's *Elements* had come to seem irrefragable. It had loomed as a monument. Now the monument seemed defaced, or if not defaced, then in some way damaged.

In the world beyond mathematics, non-Euclidean geometry represented a revolution in thought, and nineteenth-century thought was sympathetic to revolutions. It reveled in them. The details of non-Euclidean geometry were of interest only to mathematicians, of course, but its very existence provoked a certain anxiety among philosophers. "In addition to Euclidean geometry," Bertrand Russell remarked in recalling his own experiences, "there were various non-Euclidean varieties, and . . . no one knew which was right." A crack had opened in the crust of mathematics. Russell thought he saw something dreadful underneath. "If mathematics was doubtful, how much more doubtful ethics must be," he wrote. "If nothing was known, it could not be known how a virtuous life should be lived."

It is not surprising that Russell should recall that such inferences troubled his adolescence. Euclidean geometry had offered him a fixed point of certainty. Now it was gone. He felt adrift. All around him, similar inferences had gathered force. They are by no means anachronistic: they have retained their force. The impression is today widespread that if it requires anything at all, science requires an unsentimental rejection of common sense, a recusal of principles that for thousands of years have served the human race. "My own suspicion," the evolutionary biologist J. B. S. Haldane once remarked, "is that the Universe is not only queerer than we suppose, but queerer than we can suppose."

What is either common sense or common human experience when set against such exorbitant queerness?

Mathematicians had for centuries struggled to prove Euclid's parallel postulate. They were right to sense its importance. They were wrong in thinking that they could prove it. In the nineteenth century, they gave up trying. The advent of non-Euclidean geometry signaled the first in a series of percussive shocks. If the new geometries were strange, theories to come would be stranger still. The claims of common sense and common experience were assessed and then rejected:

"Thou art weighed in the balances and thou art found wanting."

ODDLY ENOUGH, THE possibilities for the development of non-Euclidean geometries should have been evident well before the nineteenth century. The captains of English, Spanish, and Portuguese ships crossing the great oceans knew perfectly well that a straight line was not necessarily the shortest distance between two points. It might have been a clue, one that is obvious today from airplane flight. On the surface of the earth, the shortest distance between two points is the arc of a great curve. Flying from sunshine to sunshine, North American passengers often find themselves staring at the snows of some hyperborean horror in between one beach and another. Geometers call the arc of flight a *geodesic*, a useful term that has come to mean the shortest distance between any two points, no matter the underlying space.

The Euclidean plane is a surface of old, and so is the earth's surface, the first familiar from textbooks, the second from life. A spherical surface envelops a sphere, two dimensions wrapped around three. When the sphere is sliced by a plane passing through its center, the plane traces a circle on its interior surface. Between any two points, the circle describes a geodesic.

The surface of the earth is a model of spherical geometry. The Euclidean blackboard is flat, but the surface of the earth is curved—and curved in the same way at every point. It is *positively* curved, a designation that unaccountably suggests a geometrical accomplishment on the order of overeating. In spherical geometry, straight lines are arcs and there are no parallel lines at all. Wandering geodesics intersect one another as they circumnavigate the globe. Lines of latitude *are* parallel, but save for the equator, they are not geodesics. The interior angles of a triangle add up to more than 180 degrees. Euclidean figures bulge as if bursting.

If Euclid is demoted on the surface of a sphere, he cannot be altogether denied. Geodesics on the surface of a sphere rest, after all, on its *surface*. The interior of a sphere remains a part of the general Euclidean background. This veil of indifference dropped, Euclidean straight lines return to prominence. The shortest distance between two sunny beaches is a straight line drilled through the earth from point to point.

The drill is a reminder. Having been cast away, Euclid has a tendency to return to any exercise in non-Euclidean geometry as an enveloping space, a contrasting structure, an astonishingly durable ghost. That ghost—is he lingering for any good reason? An ant may determine the non-Euclidean character of a sphere without *any* Euclidean contrast at her disposal. She may well conclude from purely local clues that the surface of a sphere is curved.

Clever ant.

But while the ant is looking at the surface of the sphere, who is looking at the ant, and from which perspective?

Enter ghost.

Gauss had grasped the principles of non-Euclidean geometry; he had entertained his provocative thoughts in the silence of his study. It was left to the Hungarian mathematician János Bolyai and the Russian mathematician Nikolai Ivanovich Lobachevsky to do the rest.

No story in the history of mathematics is more romantic. Bolyai's father, Farkas, had been an amateur mathematician; he was often in correspondence with Gauss. Euclid's parallel postulate obsessed him. Failure to establish the parallel postulate he regarded as "an eternal cloud on virgin truth." The proofs that he sent eagerly to Gauss, Gauss promptly returned, the errors marked. Bolyai's son, János, was a prodigy and a polyglot, the master of nine difficult languages, a mathematician of distinction, a man of many gifts. Highstrung and independent, he was consumed by duels, dances, and debts; he spent years in military service. And like his father, he was obsessed by the parallel postulate.

His father saw his son advancing toward the sinister, dark defile that had for so long obsessed him. He endeavored to warn him by means of letters that mingled the plangency and hysteria of a train's whistle: "Do not in any case have anything to do with the parallels. I know every twist and turn in this business and I have myself wandered in its fathomless night, which has extinguished every light and joy in my life. I beg you in the name of God. Leave the parallels in peace."[2]

---

2. *"Die Paralleln auf jenem Wege sollst Du nicht probieren: ich kenne auch jenen Wege bis zu Ende, auch ich habe diese bodenlose Nacht durchmessen: jedes Licht, jede Freude meines Lebens sind in ihr ausgelöscht worden. Ich beschwöre Dich bei Gott! Lass die Paralleln in Frieden."* Pretty strong stuff.

There is the rattle of thunder in all the old Hungarian clouds, a flash of lightening, claptraps accumulating, father and son receding:

Ash on an old man's sleeve
Is all the ash the burnt roses leave.
Dust in the air suspended
Marks the place where a story ended.[3]

MUCH FURTHER TO the east, a Russian mathematician, Nikolai Ivanovitch Lobachevsky, was entering the same dark defile and finding it altogether to his taste. Like Bolyai, Lobachevsky was a man of intellectual powers that had been celebrated from his youth. He was original, determined, disciplined, and hardworking. When officials at the University of Kazan discovered that he was prepared to do well any task assigned him, they assigned him every task and quickly took it for granted that he would do them well. They were not mistaken. Knowing little about architecture, Lobachevsky designed a stately and imposing university building. He became the administrative center of mathematical life, reading scholarship applications, searching out other mathemati-

3. T. S. Eliot, "Little Gidding," *Selected Poems of T. S. Eliot* (New York: Harcourt Brace Jovanovich, 1991).

cians, placating and pleasing the faculty, and skillfully managing the bureaucracy by which the university lived, its endless accountants, survey takers, patronage peddlers, censors. He took charge of the university library. Until his tenure, it had been little used and poorly managed, the mice scampering through its shelves. He bound the old books and ordered new journals; he cleaned the place up.

LOBACHEVSKY'S REJECTION OF the parallel postulate, which he published in *The Kazan Messenger* in 1829, could not have been more severe. Instead of the Euclidean plane of old, there is the hyperbolic plane of new. Both planes are planes in that they are two-dimensional surfaces. There the similarity ends. Were the hyperbolic plane like the Euclidean plane, there would be no point in denying Euclid's parallel postulate. In that way, lies madness.

Lobachevsky's illustrations of the hyperbolic plane were contrived to fit into a small region of the Euclidean plane, but beyond the margins of the illustrations, the hyperbolic plane departs from flatness, turning itself over like an inside-out orange peel. The illustrations nonetheless convey what is unusual and strange. There is a straight line R, a point *B* marked on the line, a point *P* lying beyond R; there are straight lines *x* and *y* that pass through *P*; and there is an angle θ (Figure IX.1).

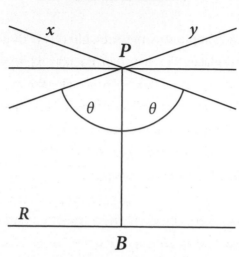

FIGURE IX.1. The hyperbolic plane

If θ is less than ninety degrees, straight lines passing through *P* will sooner or later intersect line R. If θ equals 90 degrees, a reversion to Euclid. The reversion marks the limits of the familiar.

But if θ is *greater* than 90 degrees? First, those straight lines are straight. Second, they are parallel to line R. And, third, there is *more than one* of them.

The theorems that follow are an imposition on common sense. The interior angles of a hyperbolic triangle sum to *less* than 180 degrees: A + B + C < π. Similar triangles are congruent. Lines that are parallel to a given line need not be parallel to one another. The circumference of a circle whose radius is R is *greater* than 2πR.

And Euclid's parallel postulate is false.

WITH A FEW simple straight lines, Nikolai Ivanovich Lobachevsky had placed the denial of Euclid's parallel postulate squarely on the flat Euclidean page. The theorems that he derived in his masterpiece he derived properly. They followed impeccably from the axioms of neutral geometry and the denial of Euclid's parallel postulate. The engine of inference purred without pause.

But if the denial of Euclid's parallel postulate is to be convincing, it requires more than a derivation. It demands a model and so a way of being true. Illustrations have done what illustrations can do. If no model is forthcoming, there is no reason to suppose that the denial of the parallel postulate and the axioms of neutral geometry are consistent, mutually life-enhancing. If they are not, we are returned to the logical point from which non-Euclidean geometry represented a flight. An inconsistency would indicate a contradiction between the denial of the parallel postulate and the axioms of neutral geometry.

And this is precisely what no mathematician had been able to discover.

PICTURES WHEN THEY appeared were for this reason welcomed. They revealed a world that might satisfy the axioms of non-Euclidean geometry. They were a source of inspiration and so a source of comfort.

It could be done. That is what the models suggested.

The *Kazan Messenger* had come and gone, when in 1868, the Italian mathematician Eugenio Beltrami published his influential study, "Saggio di interpretazione della geometria non-euclidea" (A study of the interpretation of non-Euclidean geometry). With the study, a model: the Beltrami pseudosphere. The pseudosphere is odd enough to prompt the suspicion that on its surface, *anything* might be true.

Illustrations of the Beltrami pseudosphere are vivid and often lovely, but they are incomplete (Figure IX.2). For one thing, the middle of the pseudosphere is girdled by what is plainly a Euclidean annulus or ring. What is that doing there? And for another thing, the *embouchure* of the pseudosphere's horns, where a divine trumpeter might have placed his lips, recedes in real life into the infinite distance. This no illustration can show.

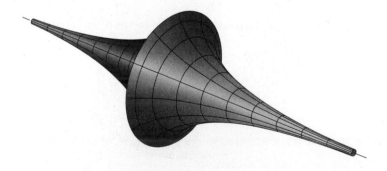

FIGURE IX.2. Beltrami pseudosphere

If illustrations of the Beltrami pseudosphere are not complete, they are nevertheless exhilarating. They show a universe coming to quicken. Shapes hidden by Lobachevsky's proofs come to life in Beltrami's picture. The triangle is an example. The Euclidean triangle is a familiar shape of art and architecture, its sides soaring or squat. When inscribed on the Beltrami pseudosphere, Euclidean triangles sag inwardly, their sides concave, and their interior angles compromised from the first by the negative curvature of their surface (Figure IX.3). The interior angles of a triangle on a pseudosphere sum to *less* than two right angles. On a sufficiently curved surface, those interior angles are lucky to sum to anything at all.

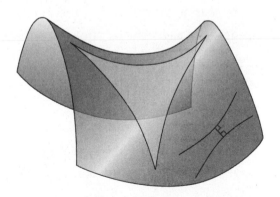

FIGURE IX.3. Hyperbolic triangle

In 1880, the superb French mathematician Henri Poincaré added another model to the growing gallery of non-Euclidean models, and in the Poincaré disk, another picture. The Beltrami pseudosphere invites the mathematician to seize it by its two horns and, perhaps, give it a toot. No one is about to seize the Poincaré disk. Among mathematicians who understood its nature, no one was tempted to go near it.

The Poincaré disk divides the Euclidean plane into three distinct regions of space. There are those points lying *beyond* the disk, those *on* its circumference, and those *in* its interior. From the outside, the disk is simply a little unit circle, as fixed and finite as a penny. From the inside, it encompasses the whole of the infinite hyperbolic plane. Outside the circle, everything is Euclidean, and inside, everything hyperbolic. Outside and inside are Euclidean from the outside, but hyperbolic from the inside. The inside is accessible from the outside—*step right in*—but not the outside from the inside—*no exit.*

There is no distinction between Euclidean and hyperbolic points. Points are points. Hyperbolic lines are otherwise. They are not like Euclidean lines at all and require careful definition. Those careful mathematical definitions express and make precise something like a dream sequence in which the Poincaré disk floats serenely in the Euclidean plane, a circle among other circles. Now and then, a drifting circle penetrates the circumference of the Poincaré disk, depositing, before it drifts on, the trace of a circumferential arc on its interior, one meeting the circumference at right angles. These Euclidean arcs are the straight lines of the hyperbolic plane. They are lines because they are lines, and they are straight because, although curved on the outside, they are straight on the inside (Figure IX.4).

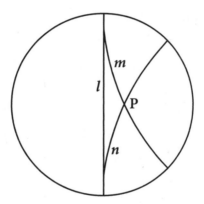

FIGURE IX.4. Poincaré disk

IN NON-EUCLIDEAN geometry, ideas that in most circumstances would fly apart stick together. This is especially true of the Poincaré disk. The binding force required to make fly-apart ideas stick together is expressed by the definition of hyperbolic distance. In the Euclidean plane, the distance $d(x, y)$ between two points x and y is the length of the straightest line joining them: $ds^2 = dx^2 + dy^2$. Square roots

are squared in this formula to preserve the positive character of distance itself. Distances cannot be negative. But Euclidean distances are unavailing inside the Poincaré disk.
Their introduction there would do nothing more than ratify
all the old Euclidean postulates and prejudices. There is, instead, hyperbolic distance, a measure that justifies, if anything does, the designation of the disk as deviant.

To define the hyperbolic distance between any two
points in the Poincaré disk requires, as one might expect,
a number of unfamiliar ideas. But some sense of deviant
distance is accessible by means of a formula that traces
only the distance from the center of the Poincaré disk to its
circumference. Once given, the full definition of distance
follows. The formula is $ds = 2dr / 1 - r^2$. The symbol ds
designates the hyperbolic distance of a radial line originat-
ing in the center of the disk and proceeding sedately to its
circumference. The symbol dr measures its *Euclidean* distance. Two different measures of distance have been assigned to one and the same line, but they are codependent.
Hyperbolic distance has been defined in terms of Euclidean distance. The question of the distance from the center
of the disk to its circumference thus receives two quite
distinct answers. Standing beyond the boundary of the
Poincaré disk, the Euclidean geometer sees that the circumference of the disk is along its radius one unit away
from its center. *How far? Not far.* But as r approaches 1,

ds approaches infinity. Standing at the center of the disk and then trudging in any direction, the hyperbolic geometer sees the circumference eternally receding. *How far? Very far.*

Differences in distances ramify throughout the hyperbolic plane. The Euclidean geometer will observe from well beyond the disk that within the disk, distances contract as his hyperbolic colleague advances on a line toward the boundary. That poor fat fool is shrinking before his eyes. This the hyperbolic geometer does not see at all. He is in his advance toward the boundary precisely as he was reposing at its center. Everything is as it was.

THE DEFINITION OF distance by which the hyperbolic world is ruled brings about the failure of Playfair's axiom. The curved lines throughout the disk are straight because they are geodesics, and they are geodesics because they represent the shortest distance between two points. The Euclidean geometer can cast a cold eye into the Poincaré disk and see why Euclid's parallel postulate fails.

THE POINT *P* lies somewhere on the disk, and two lines *m* and *n* pass through *P*, both of them approaching the circumference at two different points *A* and *B*, the place—the very place—at which AB itself passes through the circumference on its way toward strictly Euclidean glory.

The chord PA is parallel to *l* because PA and AB do not intersect *within* the Poincaré disk. How could they? In A, they have a common point of intersection at the boundary. Strictly Euclidean, that point of intersection is located beyond the interior of the disk. Considered as hyperbolic lines, both AB and PA are infinitely far from A.

Thus AB has in PA one parallel line meeting the conditions of Playfair's axiom.

But AB meets the circumference *twice*, once at A and again at B. Two lines pass through P, and *both* are parallel to AB.

POINCARÉ WAS A powerful mathematician and a subtle philosopher. He had induced the pendulum to swing wide, but he knew all about pendulums, and knew perfectly well that having swung wide, they have a tendency to swing back.

"Let us consider a certain plane," Poincaré wrote in a brilliant little book, *Science and Hypothesis*, "which I shall call the fundamental plane, and let us construct a kind of dictionary by making a double series of terms written in two columns, and corresponding each to each, just as in ordinary dictionaries the words in two languages which have the same signification correspond to one another." The words contained in this dictionary are *space, plane, line, sphere, circle, angle, distance*, and the like:

Let us now take Lobatschewsky's [sic] theorems and translate them by the aid of this dictionary, as we would translate a German text with the aid of a German-French dictionary. *We shall then obtain the theorems of ordinary geometry* [italics in original]. For instance, Lobatschewsky's [sic] theorem: "The sum of the angles of a triangle is less than two right angles," may be translated thus: "If a curvilinear triangle has for its sides arcs of circles which if produced would cut orthogonally the fundamental plane, the sum of the angles of this curvilinear triangle will be less than two right angles."

With these remarks, Poincaré invited an affronted common sense to reacquire a say in the Euclidean Joint Stock Company. The theorems of hyperbolic geometry are theorems of Euclidean geometry; they are theorems of Euclidean geometry disguised by a new, radical definition of distance. On back translation, the disguise drops away. Familiar old faces appear again. Euclidean and hyperbolic geometry are *not* two entirely different theories. They coincide in their assessment of the truth. There is amity between them.

And Euclid? What might *der Alter* have said? He might have said that finding an interpretation of distance under which the parallel postulate fails is an interesting exercise in misdirection, but one remote from his own concerns. A philosopher disposed to doubt that snow is white, on the grounds that "snow" might mean bauxite is not commonly understood to have discovered anything about snow.

Euclid might have said this.

IN 1872, THE German mathematician Felix Klein delivered a lecture at the University of Erlangen under the title "Vergleichende Betrachtungen über neuere geometrische Forschungen" (A comparative consideration of recent developments in geometry). His lecture was very much a manifesto. Klein had just joined the faculty at the university, Herr Professor and Herr Klein sharing a podium, an office, and a space on many of the same dotted lines; and their manifesto became known as the Erlangen program.

The Erlangen program was a call to classification and so a call to arms. Strange geometries were proliferating throughout European mathematical circles. Their significance, Klein argued, could never be assessed until their relationships were understood. The beloved Euclid of myth and memory was not so much demoted as absorbed, Euclidean geometry finding a place in Klein's scheme, but one place among many.

The classification that Klein imposed on the unruly world of nineteenth-century geometry was both geometrical and algebraic. The chief business of the law is, as Charles Dickens observed, to make business for itself. It is a principle not commonly observed to fail within mathematics. If there are two systems at work in the classification of various geometries, their analysis might occupy mathematicians for more than a century. And it did.

Klein's own analysis featured a reversion to certain familiars from family life in which Euclidean geometry is like a cherished and pampered son surrounded, especially on ceremonial occasions, by a curious constellation of uncles: fond, jovial, exuberant elliptical geometry; dark, scowling, saturnine hyperbolic geometry, his visits the occasion to remind everyone that family is, after all, family; and balanced, lucid, wise projective geometry, *der Onkel*, in Klein's eyes, an uncle among uncles but chief among them anyway.

Projective geometry came to its flowering in the high Renaissance as a painterly method, a way of capturing in the two dimensions of stretched canvas a world that insists on conducting its affairs in three. In the real world, railway tracks receding into the distance maintain a fixed distance from one another, but in the painter's world, they converge toward a distant, soundlessly spinning point. Chinese artists did not bother with perspective, and young children do not notice it.

The projective plane is very much like the Euclidean plane; existing in two dimensions, it stands between the human eye or the artist's canvas and an object or a landscape in three dimensions. Nineteenth-century art schools often encouraged students to master perspective by drawing directly on a flat plane of glass held directly before a scene.

THE CLASSIFICATION OF uncles is one part of the Erlangen program. The other part is, as other parts so often are, more interesting because more algebraic and so more abstract. Évariste Galois, shortly to die in some dismal duel—rival, romance, revolver—had, one hundred years before Steinitz composed his work on fields, introduced mathematicians to the greatest and most powerful of algebraic abstractions in the idea of a group. Mathematicians thereafter did what Steinitz urged them to do by separating the group from its examples. A group G is a set of objects $G = \{a, b, c, \dots\}$. These objects are closed under an associative operation, $a \circ b$. An *operation*, meaning that a is imposing itself on b. *Closed*, meaning that whatever the operation, the result is still in G. *Associative* meaning that $(a \circ b) \circ c = a \circ (b \circ c)$. There is an identity element e in the group such that for every element a, $a \circ e = a$. And for every element a, an inverse a^{-1}, such that $a \circ a^{-1} = e$. The positive and negative integers form a group under the familiar operation of addition. Whenever two integers are added together, the result is yet again an integer. Summing integers is an operation indifferent to temporal order, and so associative. It hardly matters whether 3 and 5 are first added together and then added to 12, or whether 5 and 12

---

go first, with 3 tacked on afterward. Zero is an identity element for this group. When added to any integer, it does absolutely nothing. And every integer has an inverse in its negative mirror: 5 plus −5 returns sullenly to zero.

There is nothing more, although it may, perhaps, be acknowledged that this is quite enough.

What, then, are the groups to which a geometry might be attached? This is Klein's question. What are the Euclidean groups? Ours.

Having stared so long at the Euclidean blackboard, the geometer, let us say, undertakes to penetrate its surface and move around. Once blackboard bound, he may move by translation, rotation, or reflection. The geometer is himself hardly necessary to the ideas that follow. He may be allowed decently to disappear. The idea of a permissible move remains as his mathematical trace.

The Euclidean plane comprises points and exists in two dimensions. In translation, the geometer goes from one point to another along a straight line. The translation remains as a transformation, or mapping, of the plane back on to itself, a point-to-point mapping such that the starting place is mapped to the stopping place. Everything else remains the same. Rotations and reflections are also mappings of the Euclidean plane onto itself, an abstract way of recording what the geometer has done without ever bringing about his resurrection.

These transformations, or mappings, are the elements of a group. The group operation is the succession of transformations. It is easy enough to contrive identities and inverses among the transformations. Mathematicians have been doing it for more than two hundred years.

If the transformations preserve distances, so that under their action, things that were far apart remain to the same degree far apart, the transformation is called an *isometry*, and the result is a famous old group in mathematical physics, the Euclidean group $E(n)$. It is a group with as many arms as Vishnu, describing the plane if $n = 2$, and ordinary space if $n = 3$. The transformations themselves are called Euclidean moves.

If those isometries happen to preserve orientation as well as distance, clockwise going to clockwise, and the reverse, the Euclidean group becomes the special Euclidean group, $SE(n)$. This is an important group in analytical mechanics; it describes the behavior of rigid objects. For this reason, transformations are called rigid body moves, rather a spastic designation, all things considered, and the rigid body moves are precisely the old familiars of translation, rotation, and reflection.

THE CLASSIFICATION OF various geometries by means of their groups embodied the heavy cutting edge of a large and generous research program. It embodied, as well, the last stage in an ongoing drama within Euclidean geometry. Writing so long ago, Euclid had retained a vital connection between his geometrical structures and some purely human power of getting things to move around in space. In group theory, that power is promoted to an abstract pantheon and then disappears in favor of the group's transformations, their *actions*, as mathematicians sometimes say, the link that is severed in theory returning in etymology. The promotion to abstraction is today general in Euclidean geometry: the *shapes* to Platonic forms, incomprehensible but irreplaceable; the *constructions* to derivations; the *ruler and compass* to numbers; Euclidean *motion* to transformations.

And the Euclidean Joint Stock Company?

Ownership has been diluted. There are additional members on the board. A new feeling prevails. But the old Euclidean Joint Stock Company is still known by its proper name: it is the *Euclidean* Joint Stock Company.

Chapter X

EUCLID THE GREAT

> *Whatever withdraws us from the power of our*
> *senses, whatever makes the past, the distant,*
> *or the future predominate over the present,*
> *advances us in the dignity of thinking beings.*
>
> —SAMUEL JOHNSON

CLASSICAL EUCLIDEAN GEOMETRY is, in a narrow sense, an exhausted discipline. No student of mathematics is occupied in adding to the theorems that Euclid demonstrated ones that he might have overlooked.

The sturdy old oak, having weathered so many winter storms, occasionally puts forth a few resplendent new leaves. In 1899, the American mathematician Frank Morley discovered an exquisitely beautiful theorem. At the point of intersection of three angle trisectors, there is always an equilateral triangle.

Beautiful as Morley's theorem is, much that might have been discovered by means of the Euclidean system *has* been discovered. With so little left to learn, the study of hard Euclidean problems has become something of a recreational obsession. These are problems that are easy to state but difficult to prove. The Steiner-Lehmus theorem is an example. Is any triangle whose angle bisectors are equal isosceles? Every now and then, an accomplished mathematician, having assured himself that he might knock over this problem quickly, tries to knock it over quickly, emerging days or weeks later, saying, if he is truthful, *I did it, but it almost killed me.*

TO MATHEMATICIANS, EUCLID offered a method of proof and so a way of life. That this method should have remained as an ideal for more than two thousand years is remarkable. An educated Greek, Euclid's sensibilities must have been formed by the Homeric epics, as familiar to men of his time or place as Shakespeare is to us. But if the Homeric epics survived in the vault of Greek memory, the Homeric style had declined entirely into desuetude by the time that Euclid scratched his first diagram into the dust. The manner vanished with its maker. This is surely true today. No one but a lunatic would think to compose an epic poem.

The Euclidean style endures. It is vital, an ideal, a moral advantage, a corrective to whatever is spongy, soft, indistinct, slovenly, half-hidden, half-formed, half-baked, or only half-right, the mind in full possession of its powers, straight as an arrow, hard as a stone, uncompromising as a bank. "Pre-Scientific man," observed the superbly original French mathematician René Thom, "must have had an implicit knowledge of the geometry of space and time." Prescientific man obviously knew his way around: he would not have otherwise survived. But "only with the advent of Greek geometry," Thom adds, "was this knowledge to attain an explicit, hence a *deductive* form" (italics added).[1]

In modern life, to be explicit is to be frank and thus willing to be tactless in the discussion of odious sexual details. This sense of the word is secondary. Derived from the Latin *explicare* and the French *explicite*, the word means to unfold; it carries the connotation of progressive revelation. The slow and painful undertaking by which the theorems of Euclidean geometry are derived from its axioms is an unfolding. The world of the senses recedes. The mind expands. A complex new figure emerges in thought, one expressing the relationship between the axioms of a system, its theorems, and its illustrations. The relationship cannot be seen at once; it must be understood. It is not immediate; it must be acquired. An axiomatic system is like the sonata or the nineteenth-century novel. Where the listener first hears a succession of melodies, the mathematician hears a theme

---

1. René Thom, *Semio Physics* (Reading, MA: Addison-Wesley, 1990), 32.

and its development. A sense of coherence must be earned. It cannot be granted.

And it does not come easy.

How LONG? How long is it destined to last, the severe Euclidean ideal?

It is the purpose of a proof to compel belief. Violence often works to compel assent, and if not violence, then its threat. But a belief that is not freely given cannot easily be extorted. Mathematicians know this. They have reposed their confidence in their proofs.

Writing in the December 2005 issue of the *Notices of the American Mathematical Society*, the mathematician Brian Davies saw doubt creeping into all the sacred places on its little rat's feet. In 1931, Kurt Gödel demonstrated that whole-number arithmetic is incomplete. No matter the axiomatic system of arithmetic, it had limits beyond which certain propositions of the system could not be demonstrated in the system. Gödel argued for his results by proving them. This did not by itself undermine the certainty of mathematics. Proof is, after all, proof. It did nothing to enhance the certainty of mathematics, either.

The decision by mathematicians to allow certain proofs to be completed (or verified) by the computer was undertaken in the 1970s. It has provoked skepticism ever since. In 1852, Francis Guthrie asked whether any map could be colored using just four colors to mark its distinct regions. No one knew. Simple to state, the problem is difficult to solve. In 1976, Kenneth Appel and Wolfgang Haken offered the mathematical community a proof of the four-color theorem. It demanded that a great many separate cases that could never be verified by hand be verified by the computer.

No one thought that Appel and Haken's proof was mistaken. No one was completely convinced that it was not. No one knew quite what to think. To this day, no one does.

Davies raised another point, sadder than the others and more poignant. The finite simple groups are scattered throughout mathematics. Classifying them has been a considerable project, one involving many mathematicians. Proofs now run to tens of thousands of pages, but, says a reviewer of "Whither Mathematics?" Davies's paper on this issue, "no one knows for certain whether this body of work constitutes a complete and correct proof. . . . [S]o much time has now passed that the main players who really understand the structure of the classification are dying or retiring, leaving open the possibility that there will never be a definitive answer to the question of whether the classification is true."[2]

Vulnerant omnes, ultima necant. Every hour wounds, the last one kills.

---

2. "Mathematics: The Loss of Certainty," *ScienceBlog*, 2005, review of Brian Davies, "Whither Mathematics?" *Notices of the American Mathematical Society* 52 (December 2005): 1350–1356.

If Euclidean geometry, narrowly understood, is exhausted, and if the way of life that Euclid offered is itself open to revision—the first step inevitably to rejection—wherein the peculiar and powerful claim that Euclid makes? Western art has always been a servant in a great Euclidean house. In an interview with Émile Bernard, Paul Cézanne remarked that there is "an invisible scaffolding of spheres, cones and cylinders in nature." Critics have been less successful in observing as much in the paintings of Claude Monet or J. M. W. Turner. Turner's early works and especially his drawings reflect an architectural appreciation of Euclidean forms. His mature paintings do not. About the great oils, Kenneth Clark says what is obvious: "They have no ready assonance in geometry." Looking at one and the same thing—nature, after all—each man came away with markedly different ideas about that invisible scaffolding. No very good sense can be given to the idea that the elements of Euclidean geometry may be found *in* nature because either everything is found in nature or nothing is. Euclidean geometry is a theory, and the elements of a theory may be interpreted only in terms demanded by the theory itself. Euclid's axioms are satisfied in the Euclidean plane.

Nature has nothing to do with it.

Euclid has achieved a permanent hold on the human imagination for reasons that go beyond his manner, his method, the details of his proofs, or even the many ideas he has offered the mathematical community. Beyond any other book, it is the *Elements* that has offered an uncompromising appreciation of the world of shapes—one that it created. The *Elements* is an exaltation of geometry. Euclid made a conscientious but unsuccessful effort to incorporate into his thoughts the numbers and their properties, but it is to geometry that his heart owes its allegiance. Because this is so, he was able to offer mathematicians what mathematicians so rarely offer, and that is a vision.

The vision offered, mathematicians could ask the question that only a vision could make possible: what form of unity lies beneath the numbers and the shapes—*le coeur dans le coeur*, the deepest structure, the heart of the profound identity between shapes and numbers.

> *They reckon ill who leave me out*
> *When me they fly, I am the wings*
> *I am the doubter and the doubt,*
> *And I the hymn the Brahmin sings.*[3]

Of this form of unity, we know more than Euclid could have known. The quest for unity will continue, and, of course, it will always fail. And this, too, we know. Whatever

---

3. From Ralph Waldo Emerson, "Brahma."

the form of unity mathematicians acquire, the world's diversity will in time overwhelm them, as it overwhelms us all.

Euclid remains what sensitive men and women always thought he was, a great partisan, an unequivocal voice, a part of the drama in which opposites are forever resolved and then as often dissolved.

I am writing about Euclid of Alexandria; the Euclid of the *Elements*; the Euclid of geometry, dust boards and diagrams, procedures and proofs, points and planes.

I am writing of Euclid the Great.

Teacher's Note

In his twenty-ninth proposition, Euclid says . . .

To write about Euclid is to imagine oneself linked as a companion in art to men and women long dead but still shuffling toward the Euclidean blackboard. The low murmur of their vows may be heard in Greek, Latin, or Arabic; it may be heard in all of the languages in which books are made and memories preserved.

Everyone teaching from or writing about Euclid's *Elements* does so from his own perspective, of course, but Euclidean plane geometry is not a subject that encourages pedagogical innovation. The elements are always the same: Euclid's common notions, his definitions, his axioms, and then his theorems and their proofs. There is a sense, at times subdued and at times ebullient, that this very old system merits a form of devotion, teachers and their students participating in a ritual whose full meaning is not easily grasped and never grasped at once.

The Euclidean tradition stretches from the ancient world to our own, but its value is not in the end the propositions that it makes possible. These we know and we have known them for a long time. "What can be shown," Wittgenstein remarked in the

Tractatus, "cannot be said" (*Was gezeigt werden kann, kann nichts gesagt werden*).

As much is true of Euclid's *Elements*.

The book demands both effort and concentration. The proofs do not come easy. A way of life is requested, and if it is not forthcoming, it is demanded. It is a way of life and a form of dedication that has a striking moral value. It is noble. This the *Elements* does not say, but everyone coming to the book understands that this is what it shows.

The Euclidean academy is remarkably stable. It has lasted a long time. Teachers and writers and their students enter the academy and are then lost on the sands of time. It does not matter. The academy confers a form of immortality on its academicians. It is the immortality that arises from having participated in one of the arts of civilization. This is the only form of immortality that any of us can share.

Teachers and writers alike hope that having been taught, they will be able to teach others in turn. It is a *hope*.

But no one writing about Euclid is entitled to end a book in doubt.

Now in his thirtieth proposition, Euclid says . . .

A Note on Sources

All references to Euclid are from Euclid, *The Elements: Books I–XII*, complete and unabridged, edited and translated by Thomas L. Heath (New York: Barnes & Noble, 2006). The original edition of Heath's text was published in 1906; his textual comments, although sometimes valuable, are, of course, out of date.

Appendix

Euclid's Definitions

Definitions in boldface represent the core of Euclid's scheme, the load-bearing structures.

1. **A point is that which has no part.**
2. **A line is length without breadth.**
3. **The extremities of a line are points.**
4. **A straight line is a line which lies evenly with the points on itself.**
5. **A surface is that which has length and breadth only.**
6. **The extremities of a surface are lines.**
7. **A plane surface is a surface which lies evenly with the straight lines on itself.**
8. A plane angle is the inclination to one another of two lines in a plane which meet one another and do not lie in a straight line.
9. And when the lines containing the angle are straight, the angle is called rectilineal.
10. When a straight line set up on a straight line makes the adjacent angles equal to one another, each of the equal

angles is right, and the straight line standing on the other is called a perpendicular to that on which it stands.

11. An obtuse angle is an angle greater than a right angle.

12. An acute angle is an angle less than a right angle.

13. A boundary is that which is an extremity of anything.

14. A figure is that which is contained by any boundary or boundaries.

15. A circle is a plane figure contained by one line such that all the straight lines falling upon it from one point among those lying within the figure are equal to one another;

16. And the point is called the center of the circle.

17. A diameter of the circle is any straight line drawn through the center and terminated in both directions by the circumference of the circle, and such a straight line also bisects the circle.

18. A semicircle is the figure contained by the diameter and the circumference cut off by it. And the center of the semicircle is the same as that of the circle.

19. Rectilineal figures are those which are contained by straight lines, trilateral figures being those contained by three, quadrilateral those contained by four, and multilateral those contained by more than four straight lines.

20. Of trilateral figures, an equilateral triangle is that which has its three sides equal, an isosceles triangle that which has two of its sides alone equal, and a scalene triangle that which has its three sides unequal.

21. Further, of trilateral figures, a right-angled triangle is that which has a right angle, an obtuse-angled triangle that which has an obtuse angle, and an acute-angled triangle that which has its three angles acute.

22. Of quadrilateral figures, a square is that which is both equilateral and right-angled; an oblong that which is right-angled but not equilateral; a rhombus that which is equilateral but not right-angled; and a rhomboid that which has its opposite sides and angles equal to one another but is neither equilateral nor right-angled. And let quadrilaterals other than these be called trapezia.

23. **Parallel straight lines are straight lines which, being in the same plane and being produced indefinitely in both directions, do not meet one another in either direction.**

Index

Addition, 21, 24, 103, 104, 110, 112,
  142, 143
Alexandria, library in, 5
Algebra, 7, 70–71, 91, 142
  geometrical algebra, 95, 96, 98,
    111 (*see also* Geometry:
    analytic geometry)
  straight line defined by equation,
    98, 112–113
  triumph of, 103, 105, 106
Analytical mechanics, 144
Angles, 41, 88, 119
  acute, 160, 161
  base angles of an isosceles
    triangle, 58
  and Beltrami pseudosphere, 133
  of curvilinear triangle, 139
  as equal, 26, 50–51, 52, 58, 64, 67,
    68, 73–74, 81, 89–90, 159
  obtuse, 160, 161
  rectilineal, 159
  right angles, 7, 50–51, 135, 160,
    161 (*see also* Pythagorean
    theorem)
  in spherical geometry, 125
  trisecting, 47–48, 147
Apollonius, 6
Appel, Kenneth, 151
Approximations, 4

Arabic archipelago, 70
Arab renaissance, 120
Archimedes, 6, 9
Architecture, 1, 4, 7, 64, 79, 128, 133
Arguments, 15–16, 17, 20
Aristippus, 1–2
Aristotle, 15, 16, 17, 23, 31, 45
Arithmetic, 69, 80, 92, 112
  and completeness axiom, 109
  consistency/inconsistency of, 107
  fundamental theorem of, 100–101
  geometrical axioms as
    arithmetical theorems, 114
  as incomplete, 150
  triumph of, 103
  See also Geometry: unity of
    geometry and arithmetic;
    Mathematics; Numbers
Art, 7, 79, 80, 133, 152, 155. *See also*
  Paintings
Associative laws, 105
Associative operation, 142
Assumptions, 11, 12, 27, 45, 46, 55,
  83, 119
  of existence of
    points/lines/planes, 49, 107
  hidden, 30–31
  that the parallel postulate is
    false, 120, 121

Index

Atiyah, Michael, 91
Atoms, 41–42, 43, 44
Axiomatic systems, 11, 12, 14, 149
  and arguments, 17
  new, 107
  as way of life, 9, 106, 148, 152, 156
Axioms, 45–56, 80, 90, 152
  Archimedean axiom, 108–109
  Cantor-Dedekind axiom, 100
  completeness axiom, 108, 109
  of connection, order, congruence
    and continuity, 107
  consistency/inconsistency of,
    106–107, 131
  for fields, 104–105, 112
  fifth axiom, 53–56, 118 (see also
    Axioms: Playfair's axiom;
    Parallel postulate)
  first axiom, 113–114
  first three axioms, 46, 48–49, 51,
    61, 66, 86
  fourth axiom, 50–51, 73
  fourth proposition as axiom, 27
  Hilbert's axioms, 109, 111
  interpreted in arithmetic,
    113–114
  made theorems, 46
  of neutral geometry, 131
  Playfair's axiom, 54, 55, 91, 137,
    138
  relationship between axioms
    and theorems, 12, 14, 19, 149
  as self-evident, 46
  See also Axiomatic systems

Babylonians, 8, 69
Bacon, Francis, 77
Beltrami, Eugenio, 121, 132–133
Bolyai, János, 118, 122–123, 126,
  127–128

Bolyai, Farkas, 127–128
Boole, George, 23
Boundaries, 160
Breadth, 33, 35, 36, 159
Bridge of Asses, 64, 65(fig.). See
  also Propositions: fifth
  proposition

Calculus, 39, 94
  differential calculus, 41, 59
  of segments, 110
Cantor, Georg, 93
Cantor-Dedekind axiom, 100, 101,
  109
Cardioids, 99, 99(fig.)
Cathedrals, 64
Causality, 13
Cézanne, Paul, 152
Change, 43, 44, 52
Chesterton, G. K., 11
China, 9
Cicero, 1
Circles, 7, 13, 25, 46, 79
  center/circumference of, 98,
    130, 135, 136, 160
  diameter of, 160
  and geodesics, 125
  and proposition one, 61–62
  radii of, 49, 62, 98, 130
  semicircles, 160
  See also Poincaré, Henri:
    Poincaré disk
Clark, Kenneth, 78, 152
Clay tablets, 8
Coincidence, 21, 23, 25–26, 27, 39,
  41, 67
  and concrete vs. abstract models
    of geometry, 28–29
Common beliefs/notions, 19–32, 90
  fifth, 29–30

first, 24
fourth, 23
second/third, 24, 62, 74
Common sense, 36, 64, 91, 118, 124, 130, 139
Commutative laws, 105
Compass, 63. *See also* Straight-edge and compass
Complexity, 55, 107
Computers, 150
Congruence, 26, 39, 67, 73, 74, 75, 107, 130
Consistency/inconsistency, 106–107, 131
Contradictions, 17, 83, 87, 89, 100, 120, 121, 131. *See also* *Reductio ad absurdum*
Contrapositives, 83, 84(n), 86, 86(fig.)
Converse relationship, 69, 81(n), 82, 83
Coordinate Method, The (Gelfand, Glagoleva, and Kirillov), 99–100
Coordinate systems, 97, 97(fig.), 115
Critique of Pure Reason, The (Kant), 117
Cultures, 3, 4, 9
Curvature, 38, 99, 125, 139
extrinsic, 40, 41
negative, 133
and straight lines, 39

Das Kontinuum (Weyl), 44
Davies, Brian, 150, 151
De Architectura (Vitruvius Pollio), 1–2
Dedekind, Richard, 102. *See also* Cantor-Dedekind axiom

Deduction, 45, 149
Definitions, 20, 33–44, 51, 90, 159–161
eighth and ninth, 51–53
fifteenth, sixteenth, and seventeenth, 62
fifth, 35
first seven and twenty-third, 33–34
fourth, 38
of hyperbolic lines/distance, 134–135, 136, 139
nineteenth, 60, 84, 85–86
ninth through twenty-second, 34
and real ordered fields, 113
of rectilinear figures, 60
seventh, 38
of shape, 49
tenth, 73
third, 43
twentieth, 60
twenty-third, 37–38, 43, 84
Degrees of freedom, 37
Democritus, 41, 42, 44
De Morgan, August, 84(n)
Descartes, René, 45, 96
Dieudonné, Jean, 115
Dimensions, 35, 36, 37, 40, 70, 125, 141, 144
Distance, 23, 37, 39, 41, 56, 69–70, 87, 88, 125, 132, 144
hyperbolic distance, 135–137, 139
Distributive laws, 105
Division, 93, 95, 103, 104, 110, 112

Egyptians, 11
Einstein, Albert, 118
Elementary Geometry from an Advanced Standpoint (Moise), 94

Elements (Euclid), 9, 44, 80, 90, 91, 123, 153
  Book I, 6, 119
  Book II, 6
  Book V, 7, 93–94, 110
  Books V through IX, 7
  Book VII, 93, 110
  Book X, 93–94, 100, 110
  books in, 6–7
  first four books, 7
  as having limited symbolic reach, 71
  as illustrated, 59, 64–65, 79–80, 87, 90
  and mountain-climbing pastoral, 57–58
  as textbook, 5–6, 7, 155–156
Eliot, George, 45
Ellipses, 13, 98–99
Empson, William, 57–58
Encyclopedia Britannica, 28
Equality, 21, 22–25, 26, 36, 62, 63, 148
  definition of, 25
  "less than or equal to," 105
  of right angles, 50–51
  of squares, 75
  transitivity of, 24
  See also Angles: as equal
Equator, 125
Erlangen program, 140, 142
Ethics, 123
Euclid, 21–22, 43, 89, 140, 145, 152–153
  and Aristotle, 15, 17
  birth/death of, 5
  double insight of, 12
  Euclidean ideal, 150
  Euclidean style, 148–149

Euclidean tradition, 155–156
  and fifth axiom (parallel postulate), 54–55, 118–119, 139–140
  as a mathematician, 6
  modern versions of, 8
  predecessors of, 6
  as a teacher, 5–6, 26–27, 79
  translations of, 8
  and unity beneath diversity of experience, 11
Euclides ab omni naevo vindicatus (Saccheri), 121
Eudoxus, 6, 94, 108
Explicit (word), 149

Fields, 103–106, 112, 113, 118, 142
Flatness, 38–39, 40–41
Flaubert, Gustave, 1
Forms (Platonic), 13, 60, 145
Four-color theorem, 151
Fractions. *See under* Numbers
Friedman, Harvey, 46

Galois, Évariste, 142
Gauss, Carl Friedrich, 41, 48, 92, 93, 118, 122, 126, 127
Gelfand, I. M., 99
Geodesics, 125, 126, 137
Geometry, 5, 6, 12, 80, 83, 112
  analytic geometry, 96–97, 98–100, 108, 109, 110, 115
  classification of geometries, 140–142
  concrete vs. abstract models of, 13–14, 28–29
  differential geometry, 41
  elliptical geometry, 141

Index

Euclidean geometry as first
theory, 108, 152
hyperbolic geometry, 139, 141
neutral geometry, 122, 131
new axiom system for, 107
non-Euclidean geometries, 8,
106, 118, 121, 123, 124–141
projective geometry, 141
revising Euclidean geometry,
51–52
solid geometry, 7
spherical geometry, 125
unity of geometry and
arithmetic, 69, 71, 91, 92, 95,
110, 111, 153–154
Geometry, Euclid and Beyond
(Hartshorne), 47
Glagoleva, E. G., 99
Gödel, Kurt, 150
Greeks (ancient), 8, 14, 15, 120, 148
Groups, 142–145, 151
Grundlagen der Geometrie (Hilbert),
106, 107, 108, 110, 111
Guthrie, Francis, 150–151

Hadamard, Jacques, 27
Haken, Wolfgang, 151
Haldane, J. B. S., 124
Hardy, G. H., 77
Hartshorne, Robin, 47
Haytham, Ibn al, 120
Hilbert, David, 13, 27, 34, 52, 80,
106–115
Homeric epics, 148
Horace, 39
Hyperbola, 99
Hyperbolic plane. *See under* Planes
Hypotenuse, 69, 100. *See also*
Pythagorean theorem

Identity, 25, 50–51, 68, 142, 144
of a point and pair of numbers,
112 (*see also* Points: point as
pair of numbers)
between shapes and numbers,
153
Inference, 15, 17, 19, 43, 44, 49,
123
rules of inference, 90
Infinite regress, 32
Infinity, 38, 49, 87, 132, 134, 137
natural numbers as potentially
infinite, 92–93
Intuition, 22, 45, 53, 54, 82
Inverse relationship, 81(n), 82, 83,
105, 142, 143, 144
Isometry, 144

James, Henry, 117
Johnson, Samuel, 33, 147
Joyce, D. E., 62
Judt, Tony, 4
Jupiter and Antiope (painting),
77–78, 79, 82, 87

Kant, Immanuel, 117
Kazan, University of, 128–129
Kazan Messenger, The, 129, 132
Kirillov, A. A., 99
Klein, Felix, 140
Kline, Morris, 34

La Géométrie (Descartes), 96
Latitude/longitude, 3
Leçons de géométrie élémentaire
(Hadamard), 27
Length, 22–23, 33, 35, 36, 159
Libri Decem (Vitruvius Pollio), 1
Lines, 79, 111

Lines *(continued)*
  curved lines, 137 (*see also*
    Curvature)
  existence of, 107
  hyperbolic lines, 134, 135
  line segments, 95, 110–111
  parallel lines, 34, 84, 84(n), 88,
    89–90, 125, 130, 138, 161
  straight lines, 7, 13, 14, 22, 23,
    25, 33, 34, 37, 38, 39, 43, 46,
    48, 50, 52, 53, 60, 61, 62, 63,
    66, 73, 80–81, 95, 98, 112, 113,
    135, 137, 143, 159, 160, 161
  straight lines as ratio of three
    numbers, 112, 113
Lobachevsky, Nicolai, 118,
    122–123, 126, 128–131, 133,
    139
Logic, 2, 12, 23, 34, 53, 54, 59, 65,
    80, 82, 83, 90, 107, 108, 119
  of relationships, 24
  See also Syllogisms

Magnitudes, 7, 94
Mallory, George, 58–59
*Mathematical Thought from
    Ancient to Modern Times*
    (Kline), 34
Mathematics, 2–3, 7, 12, 41, 83,
    151
  as doubtful, 123
  mathematical physics, 144
  and mountain-climbing
    pastoral, 57
Measurements/mensuration, 11
Middle Ages, 80
Mirror images, 68
Models, 13–14, 108
Modus ponens, 17, 82
Moise, Edwin, 94

Monet, Claude, 152
Morality, 58, 156
Mordell, Louis Joel, 57
Morley, Frank, 147
Motion, 25, 26, 27, 28, 29, 63, 143
  as impossible, 43
  power of geometrical objects to
    move or be moved, 36, 37, 39,
    52, 68, 95, 145
  rigid body moves, 144
  ways of moving in a plane, 37
Mountain-climbing pastoral,
    57–58
Mount Everest, 58
Multiplication, 103, 104, 110, 112

Newton, Isaac, 47
Non-Euclidean geometries. *See
    under* Geometry
Nothing, 41, 42, 43–44
*Notices of the American
    Mathematical Society*, 150
Numbers, 3, 7, 12, 17, 29, 30, 69,
    145, 153
  and distances, 23
  fractions, 38, 94, 102, 103
  geometrical properties of
    numerals, 92
  greatest/least numbers, 109
  identifying points in space, 36
  irrational numbers, 102
  natural numbers, 91–92, 92–93,
    95, 101
  natural numbers as potentially
    infinite, 92–93
  negative numbers, 101–102,
    103, 142
  new numbers, 101–102
  number as multitude composed
    of units, 93

and points, 109 (*see also* Points: point as pair of numbers)
prime numbers, 100
rational numbers, 94, 109
real numbers, 94, 103, 105–106, 109, 110, 111, 112
Roman numerals, 4
sets of numbers, 111–112
squaring/square roots of, 70, 72, 100, 101, 102, 103, 110, 135–136
zero, 101, 103, 104, 143

Oblongs, 161
Omar Khayyám, 120–121
On Nature (Parmenides), 42

Paintings, 77–79, 140
Pappus, 68
Papyrus, 8
Parabola, 98
Paradoxes, 38
Parallelism, 53, 56, 74, 74(n), 81, 87. *See also* Lines: parallel lines; Parallel postulate
Parallelograms, 74, 74(n), 75
Parallel postulate, 81(n), 117–124
denial/failure of, 118, 120, 123, 131, 137, 139–140
and Pythagorean theorem, 119
See also Axioms: fifth axiom
Parmenides, 42–43, 44
Parts, 34–35, 42
whole as greater than the part, 21, 29–30
Pasch, Moritz, 34
Peirce, C. S., 23
Perspective (in paintings), 141–142
Peyrard, François, 8

Planes, 14, 33, 38, 39, 40, 41, 94, 96–97, 108, 111, 112, 138, 143, 144, 152
defined, 35–36, 159
degrees of freedom of, 37
existence of, 107
hyperbolic plane, 129–130, 130(fig.), 134, 135, 137
projective plane, 141–142
Plato, 5, 13, 60, 95, 145
Playfair, Francis, 53–54. *See also* Axioms: Playfair's axiom
Poincaré, Henri, 134
dictionary of, 138–139
Poincaré disk, 134–138, 135(fig.)
Points, 3, 7, 13, 33, 37, 53, 87, 111
vs. atoms, 42, 43
"between two points," 14, 41, 43, 44, 46, 48, 50, 61, 62, 70, 95, 124–125, 126, 135, 137
and continuity, 44
defined, 34, 35, 159
existence of, 49, 107, 109
hyperbolic points, 134
point as pair of numbers, 97–98, 100, 112, 113–114, 114–115
Polygons, 48
Postulates, 12. *See also* Axioms
Praxinoscopes, 78
Precision, 4, 59
Premises, 15–16, 90
Principia (Newton), 47
Proclus, 119
Proofs, 12, 17, 19, 20, 26, 31, 47, 58, 59, 87, 150
as artifacts, 32
and common beliefs, 20, 21
as difficult, 65, 89, 148
of four-color theorem, 151
by Lobachevsky, 133

Index

Proofs *(continued)*
of parallel postulate, 119,
120–122, 124
proof by contradiction, 83 (*see
also Reductio ad absurdum*)
of Pythagorean theorem, 71–75,
96
steps in, 59
of twenty-seventh proposition,
83–87, 90
as way of life, 148, 156 (*see also*
Axiomatic systems: as way of
life)
Proportions, 7, 94, 108
Propositions, 6–7, 11, 17, 90
difficulty of, 89
fifth proposition, 58, 63–68
first proposition, 60–63, 61(fig.)
first twenty-eight propositions,
122
forty-seventh proposition,
68–75
forty-sixth proposition, 73
fourth proposition, 26–27, 36,
39, 67, 68, 74
sixteenth proposition, 83–84,
84(fig.), 84(n), 85(fig.), 86
third proposition, 66
thirty-second proposition, 119
twenty-ninth proposition, 118,
155
twenty-seventh proposition,
80–90, 81(fig.), 84(n), 86(fig.)
Pseudosphere, 132–133, 132(fig.)
Ptolemy I, 5
Ptolemy Soter, 58, 119
Pyramids, 11, 12
Pythagoreans, 12, 100
Pythagorean theorem, 68–75,
72(fig.), 100

algebraic equation of, 96
and parallel postulate, 119

Quadrilateral figures, 160, 161

Railroads, 4, 141
Ratios, 94, 100, 101, 112–113
Rectangles, 7, 96
Rectilinear figures, 60, 160
Reductio ad absurdum, 77,
83–87
Reflection (in planes), 37, 68,
143, 144
Relativity and Geometry (Torretti),
39
Relativity theory, 118
Renaissance, 8, 141. *See also* Arab
renaissance
Rhombus/rhomboid, 161
Riemann, Bernhard, 118
Rigid objects, 144
Roman empire, 3–4
Roman numerals, 4
Rotation, 37, 143, 144
Rubiyat of Omar Khayyám, 120–121
Rulers. *See* Straight-edge and
compass
Ruskin, John, 78
Russell, Bertrand, 7, 27, 28, 123

Saccheri, Girolomo, 121
"Saggio di interpretazione della
geometria non-euclidea"
(Beltrami), 132
Science, 124
Science and Hypothesis (Poincaré),
138
Self-evidence, 46, 55
Shapes, 3, 7, 12–13, 52, 60, 71, 107,
117, 133, 145, 153

coincidence of, 25–26, 36, 39
definition of, 49
Size, 22
Socrates, 15. *See also* Plato
Some Versions of Pastoral
(Empson), 57–68
Space(s), 4, 8, 12, 20, 28, 35, 37, 43,
56, 70, 125, 149
homogeneity of, 52
three-dimensional, 40, 70,
144
unbounded vs. infinite, 38
Spheres, 38, 39. *See also* Surfaces:
surface of a sphere
Square roots. *See* Numbers:
squaring/square roots of
Squares, 7, 72(fig.), 75, 79, 96, 161
St. Vincent Millay, Edna, 19
Stability, 31
Steiner-Lehmus theorem, 148
Steinitz, Ernst, 103–104
Straight-edge and compass, 47–48,
63, 145
Subtraction, 21, 24, 67, 103, 104,
110, 112
Superposition, 23, 39. *See also*
Coincidence
Surfaces, 33, 133, 159
surface of a sphere, 38, 40,
125–126
Syllogisms, 15–16

Theaetetus, 6
Theorema Egregium (Gauss), 41
Theorems, 25, 79, 90, 147
as being made axioms, 46
forty-first theorem, 74
four-color theorem, 151
of hyperbolic geometry, 139
of Lobachevsky, 131, 139

relationship between axioms
and theorems, 12, 14, 19, 149
Steiner-Lehmus theorem, 148
See also Pythagorean theorem
Theories, 112, 118
of Euclidean and hyperbolic
geometry, 139
Euclidean geometry as first
theory, 108, 152
Thom, René, 95, 107, 110, 149
Time, 4, 12–13, 28, 78, 87, 88, 142,
149
flow of time vs. points used to
mark, 44
and twenty-seventh
proposition, 81–82
Torretti, Roberto, 39
Transformations, 143–144, 145
Translation (in planes), 37, 143,
144
Triangles, 7, 13, 25, 28, 39, 79, 84,
119
and Beltrami pseudosphere, 133
curvilinear, 139
defined, 34, 160, 161
as equal, 26, 67
equilateral, 60–63, 147, 160
hyperbolic, 130, 133(fig.), 139
isosceles, 58, 64–68, 148, 160
Platonic, 60 (*see also* Forms)
right triangle, 68 (*see also*
Pythagorean theorem)
scalene, 160
in spherical geometry, 125
Trilateral figures, 160, 161
Truth, 16, 25, 117, 127, 139
Turner, J. M. W., 152

"Uber den Zahlbegriff" (Hilbert),
107

Index

Unity/diversity of experience, 11, 154

Vector spaces, 114
"Vergleichende Betrachtungen über neuere geometrische Forschungen" (Klein), 140
Vermeer, Johannes, 79
View of Delft (painting), 79
Vitruvius Pollio, Marcus, 1–2
Void, 42, 44

Voltaire, 57

Watteau, Antoine, 77, 79, 82, 87
Weyl, Hermann, 44
"Whither Mathematics?" (Davies), 151
Whymper, Edward, 57
Wittgenstein, Ludwig, 155–156

Zeno the Eleatic, 38
Zero. *See under* Numbers